CLEP College Algebra Workbook

Essential Learning Math Skills Plus Two College Algebra Practice Tests

By

Michael Smith & Reza Nazari

CLEP College Algebra Workbook

Published in the United State of America By

The Math Notion

Web: WWW.MathNotion.Com

Email: info@Mathnotion.com

Copyright © 2020 by the Math Notion. All rights reserved. No part of this publication may be reproduced, stored in a retrieval system, or transmitted in any form or by any means, electronic, mechanical, photocopying, recording, scanning, or otherwise, except as permitted under Section 107 or 108 of the 1976 United States Copyright Ac, without permission of the author.

All inquiries should be addressed to the Math Notion.

ISBN: 978-1-63620-004-0

About the Author

Michael Smith has been a math instructor for over a decade now. He holds a master's degree in Management. Since 2006, Michael has devoted his time to both teaching and developing exceptional math learning materials. As a Math instructor and test prep expert, Michael has worked with thousands of students. He has used the feedback of his students to develop a unique study program that can be used by students to drastically improve their math score fast and effectively.

- **– SAT Math Practice Book**
- **– ACT Math Practice Book**
- **– PSAT Math Practice Book**
- **– Algebra Math Practice Books**
- **– Common Core Math Practice Books**
- **–many Math Education Workbooks, Exercise Books and Study Guides**

As an experienced Math teacher, Mr. Smith employs a variety of formats to help students achieve their goals: He tutors online and in person, he teaches students in large groups, and he provides training materials and textbooks through his website and through Amazon.

You can contact Michael via email at:

info@Mathnotion.com

Prepare for the CLEP College Algebra with a Perfect Workbook!

CLEP College Algebra Workbook is a learning workbook to prevent learning loss. It helps you retain and strengthen your Math skills and provides a strong foundation for success. This CLEP College Algebra book provides you with a solid foundation to get ahead starts on your upcoming CLEP College Algebra Test.

CLEP College Algebra Workbook is designed by top math instructors to help students prepare for the CLEP College Algebra course. It provides students with an in-depth focus on the CLEP College Algebra concepts. This is a prestigious resource for those who need extra practice to succeed on the CLEP College Algebra test.

CLEP College Algebra Workbook contains many exciting and unique features to help you score higher on the CLEP College Algebra test, including:

- Over 2,500 CLEP College Algebra Practice questions with answers
- Complete coverage of all Math concepts which students will need to ace the CLEP College Algebra test
- Two CLEP College Algebra practice tests with detailed answers
- Content 100% aligned with the latest CLEP College Algebra courses

This Comprehensive Workbook for the CLEP College Algebra is a perfect resource for those CLEP College Algebra takers who want to review core content areas, brush-up in math, discover their strengths and weaknesses, and achieve their best scores on the CLEP College Algebra test.

WWW.MathNotion.COM

... So Much More Online!

✓ FREE Math Lessons

✓ More Math Learning Books!

✓ Mathematics Worksheets

✓ Online Math Tutors

For a PDF Version of This Book

Please Visit WWW.MathNotion.com

Contents

Chapter 1: Review the Basics 11
    Adding and Subtracting Integers 12
    Multiplying and Dividing Integers 13
    Order of Operations 14
    Integers and Absolute Value 15
    Sets 16
    Scientific Notation 17
    Powers of Products and Quotients 18
    Negative Exponents and Negative Bases 19
    Answers of Worksheets – Chapter 1 20

Chapter 2: Radicals Expressions 23
    Square Roots 24
    Simplifying Radical Expressions 25
    Multiplying Radical Expressions 26
    Simplifying Radical Expressions Involving Fractions 27
    Adding and Subtracting Radical Expressions 28
    Answers of Worksheets – Chapter 2 29

Chapter 3: Equations and Inequalities 31
    One–Step Equations 32
    Multi–Step Equations 33
    Graphing Single–Variable Inequalities 34
    One–Step Inequalities 35
    Multi-Step Inequalities 36
    The Distributive Property 37
    Systems of Equations 38
    Systems of Equations Word Problems 39
    Systems of 3 Variable Equations 40
    Answers of Worksheets – Chapter 3 41

Chapter 4: Linear Functions 45
    Relation and Functions 46

Finding Slope..47
Graphing Lines Using Line Equation 48
Writing Linear Equations..49
Graphing Linear Inequalities 50
Write an Equation from a Graph...51
Rate of change 52
x and y intercepts...52
Slope–intercept Form 53
Point–slope Form 54
Graphing Lines of Equations 55
Equation of Parallel or Perpendicular Lines 56
Equations of Horizontal and Vertical Lines 57
Graphing Absolute Value Equations 58
Answers of Worksheets – Chapter 4 59

Chapter 5: Monomials and polynomials 65
GCF of Monomials 66
Factoring Quadratics 67
Factoring by Grouping 68
GCF and Powers of Monomials 69
Writing Polynomials in Standard Form 70
Simplifying Polynomials 71
Adding and Subtracting Polynomials 72
Multiplying a Polynomial and a Monomial 73
Multiplying Binomials 74
Factoring Trinomials...75
Operations with Polynomials 76
Answers of Worksheets – Chapter 5..77

Chapter 6: Functions Operations and Quadratic 83
Evaluating Function 84
Adding and Subtracting Functions 85
Multiplying and Dividing Functions 86
Composition of Functions 87
Quadratic Equation 88

Solving Quadratic Equations 89
Quadratic Formula and the Discriminant 90
Quadratic Inequalities 91
Graphing Quadratic Functions 92
Domain and Range of Radical Functions 93
Solving Radical Equations 94
Answers of Worksheets – Chapter 6 95

Chapter 7: Rational Expressions 100
Simplifying and Graphing Rational Expressions 101
Adding and Subtracting Rational Expressions 102
Multiplying and Dividing Rational Expressions 103
Solving Rational Equations and Complex Fractions 104
Answers of Worksheets – Chapter 7 105

Chapter 8: Matrices 107
Adding and Subtracting Matrices 108
Matrix Multiplication 109
Finding Determinants of a Matrix 110
Finding Inverse of a Matrix 111
Matrix Equations 112
Answers of Worksheets – Chapter 8 113

Chapter 9: Sequences and Series 115
Arithmetic Sequences 116
Geometric Sequences 117
Comparing Arithmetic and Geometric Sequences 118
Finite Geometric Series 119
Infinite Geometric Series 120
Answers of Worksheets – Chapter 9 121

Chapter 10: Complex Numbers 125
Adding and Subtracting Complex Numbers 126
Multiplying and Dividing Complex Numbers 127
Graphing Complex Numbers 128
Rationalizing Imaginary Denominators 129
Answers of Worksheets – Chapter 11 130

Chapter 11: Logarithms 131
Rewriting Logarithms 132
Evaluating Logarithms 133
Properties of Logarithms 134
Natural Logarithms 135
Exponential Equations and Logarithms 136
Solving Logarithmic Equations 137
Answers of Worksheets – Chapter 11 138

Chapter 12: Conic Sections 141
Equation of a Parabola 142
Focus, Vertex, and Directrix of a Parabola 143
Standard Form of a Circle 144
Equation of Each Ellipse 145
Hyperbola in Standard Form 146
Conic Sections in Standard Form 147
Answers of Worksheets – Chapter 12 148

Chapter 13: Statistics and Probability 151
Probability Problems 152
Factorials 153
Combinations and Permutations 154
Answers of Worksheets – Chapter 13 155

CLEP College Algebra Tests Review 157
CLEP College Algebra Test Answer Sheets 159
CLEP College Algebra Practice Test 1 161
CLEP College Algebra Practice Test 2 179

Answers and Explanations 195
Answer Key 195
Practice Test 1 197
Practice Test 2 211

Chapter 1:
Review the Basics

Topics that you will practice in this chapter:

- ✓ Adding and Subtracting Integers
- ✓ Multiplying and Dividing Integers
- ✓ Order of Operations
- ✓ Integers and Absolute Value
- ✓ Sets
- ✓ Scientific Notation
- ✓ Powers of Products and Quotients
- ✓ Negative Exponents and Negative Bases

"Whenever there is number, there is beauty." –Proclus

Adding and Subtracting Integers

✏️ **Find each sum.**

1) $15 + (-35) =$

2) $(-28) + (-29) =$

3) $19 + (-27) =$

4) $57 + (-64) =$

5) $(-14) + (-19) + 64 =$

6) $54 + (-36) + 19 =$

7) $46 + (-30) + (-33) + 29 =$

8) $(-40) + (-70) + 28 + 55 =$

9) $60 + (-65) + (83 - 72) =$

10) $49 + (-55) + (90 - 67) =$

✏️ **Find each difference.**

11) $(-32) - (-7) =$

12) $40 - (-12) =$

13) $(-60) - 56 =$

14) $27 - (-17) =$

15) $58 - (76 - 29) =$

16) $19 - (-14) - (-22) =$

17) $(39 + 15) - (-46) =$

18) $49 - 17 - (-13) =$

19) $85 - 45 - (-18) =$

20) $78 - (-35) - (-63) =$

21) $89 - (-11) - (-26) =$

22) $(19 - 50) - (-95) =$

23) $46 - 49 - (-87) =$

24) $120 - (98 + 24) - (-38) =$

25) $112 - (-102) + (-81) =$

26) $108 - (-42) + (-89) =$

Multiplying and Dividing Integers

✎ **Find each product.**

1) $(-7) \times (-9) =$

2) $(-5) \times 6 =$

3) $10 \times (-15) =$

4) $(-9) \times (-25) =$

5) $(-7) \times (-12) \times 13 =$

6) $(15 - 4) \times (-11) =$

7) $25 \times (-4) \times (-5) =$

8) $(85 + 10) \times (-11) =$

9) $12 \times (-19 + 12) \times 5 =$

10) $(-15) \times (-18) \times (-20) =$

✎ **Find each quotient.**

11) $85 \div (-5) =$

12) $(-90) \div (-15) =$

13) $(-121) \div (-11) =$

14) $99 \div (-33) =$

15) $(-114) \div 2 =$

16) $(-208) \div (-16) =$

17) $198 \div (-11) =$

18) $(-364) \div (-14) =$

19) $255 \div (-15) =$

20) $(-378) \div (18) =$

21) $(-184) \div (-8) =$

22) $-437 \div (-23) =$

23) $(-570) \div (-19) =$

24) $480 \div (-32) =$

25) $(-546) \div (-21) =$

26) $(486) \div (-54) =$

Order of Operations

✏️ **Evaluate each expression.**

1) $7 + (5 \times 8) =$

2) $16 - (6 \times 9) =$

3) $(17 \times 5) + 12 =$

4) $(24 - 12) - (11 \times 4) =$

5) $35 + (18 \div 3) =$

6) $(27 \times 3) \div 3 =$

7) $(88 \div 4) \times (-5) =$

8) $(9 \times 9) + (86 - 52) =$

9) $78 + (5 \times 12) + 14 =$

10) $(60 \times 4) \div (4 + 2) =$

11) $(-15) + (14 \times 4) + 18 =$

12) $(14 \times 5) - (56 \div 7) =$

13) $(7 \times 9 \div 3) - (32 + 21) =$

14) $(45 + 11 - 14) \times 2 - 15 =$

15) $(40 - 18 + 20) \times (75 \div 3) =$

16) $75 + (54 - (45 \div 9)) =$

17) $(12 + 15 - 24) + (44 \div 4) =$

18) $(78 - 19) + (27 - 10 + 7) =$

19) $(18 \times 3) + (17 \times 9) - 52 =$

20) $65 + 17 - (45 \times 2) + 40 =$

CLEP College Algebra Workbook

Integers and Absolute Value

✎ **Write absolute value of each number.**

1) $|-19| =$

2) $|-32| =$

3) $|-50| =$

4) $|31| =$

5) $|57| =$

6) $|-76| =$

7) $|42| =$

8) $|101| =$

9) $|28| =$

10) $|-49| =$

11) $|-13|$

12) $|78| =$

13) $|100| =$

14) $|0| =$

15) $|-105| =$

16) $|-77| =$

17) $88 =$

18) $|-29| =$

19) $|112| =$

20) $|-120| =$

✎ **Evaluate the value.**

21) $|-5| - \frac{|-40|}{8} =$

22) $18 - |4 - 19| - |-15| =$

23) $\frac{|-72|}{9} \times |-9| =$

24) $\frac{|6 \times (-8)|}{3} \times \frac{|-21|}{7} =$

25) $|5 \times (-9)| + \frac{|-110|}{11} =$

26) $\frac{|-96|}{12} \times \frac{|-27|}{9} =$

27) $|-19 + 12| \times \frac{|-12 \times 13|}{7}$

28) $\frac{|-19 \times 6|}{3} \times |-11| =$

WWW.MathNotion.Com

Sets

Given A = {1, 2, 3, 8, 12}, B = {2, 4, 5, 7}, and C = {5, 7, 9, 11}, find:

1) A ∪ B _____

2) A ∪ C _____

3) B ∪ C _____

4) A ∩ B _____

5) A ∩ C _____

6) B ∩ C _____

7) (A ∪ B) ∪ C _____

8) (A ∪ B) ∩ C _____

9) (A ∩ B) ∩ C _____

10) (B ∪ C) ∩ A _____

Refer to the diagram below to find each set.

11) A ∪ B _____

12) A ∪ C _____

13) B ∪ C _____

14) A ∩ B _____

15) A ∩ C _____

16) B ∩ C _____

17) (A ∪ B) ∪ C _____

18) (A ∪ B) ∩ C _____

19) (A ∩ B) ∩ C _____

20) (B ∪ C) ∩ A _____

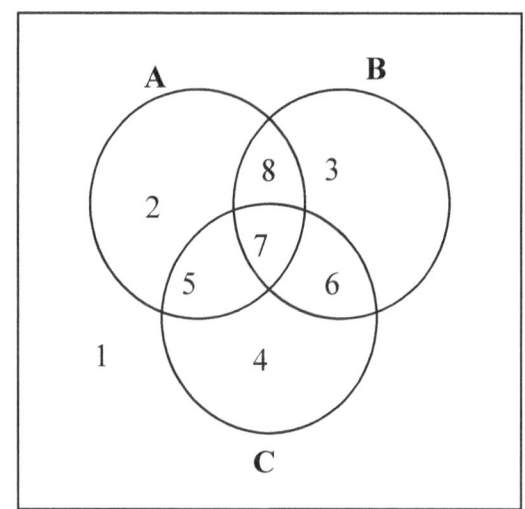

Scientific Notation

✎ Write each number in scientific notation.

1) $0.226 =$

2) $0.05 =$

3) $4.8 =$

4) $90 =$

5) $120 =$

6) $0.123 =$

7) $82 =$

8) $5,400 =$

9) $2,460 =$

10) $75,300 =$

11) $61,000,000 =$

12) $0.00009 =$

13) $468,000 =$

14) $0.00458 =$

15) $0.000087 =$

16) $31,800,000 =$

17) $950,000 =$

18) $9,000,000,000 =$

19) $0.0007 =$

20) $0.00041 =$

✎ Write each number in standard notation.

21) $4 \times 10^{-2} =$

22) $7 \times 10^{-4} =$

23) $4.3 \times 10^6 =$

24) $7 \times 10^{-4} =$

25) $8.7 \times 10^{-3} =$

26) $12 \times 10^5 =$

27) $35 \times 10^3 =$

28) $1.89 \times 10^5 =$

29) $13 \times 10^{-6} =$

30) $7.3 \times 10^{-4} =$

Powers of Products and Quotients

Simplify.

1) $(4^2)^3 =$

2) $(5^2)^2 =$

3) $(3 \times 3^2)^3 =$

4) $(3 \times 2^3)^2 =$

5) $(15^2 \times 15^2)^5 =$

6) $(7^2 \times 7^3)^4 =$

7) $(9 \times 9^2)^2 =$

8) $(4^6)^3 =$

9) $(7x^7)^3 =$

10) $(8x^4y^3)^2 =$

11) $(3x^3y^2)^4 =$

12) $(4x^2y^2)^2 =$

13) $(3x^5y^2)^3 =$

14) $(4x^3y^2)^3 =$

15) $(2x^3x)^5 =$

16) $(6x^4x^2)^2 =$

17) $(7x^{12}y^5)^2 =$

18) $(5x^7x^4)^3 =$

19) $(8x^2 \times 6x)^2 =$

20) $(9x^{14}y^3)^3 =$

21) $(5x^4y^2)^4 =$

22) $(3x^3y^7)^5 =$

23) $(8x \times 2y^3)^2 =$

24) $\left(\frac{8x}{x^3}\right)^3 =$

25) $\left(\frac{x^4y^5}{x^3y^5}\right)^7 =$

26) $\left(\frac{36xy}{6x^5}\right)^2 =$

27) $\left(\frac{x^4}{x^5y^2}\right)^3 =$

28) $\left(\frac{xy^2}{x^3y^8}\right)^{-3} =$

29) $\left(\frac{5xy^7}{x^2}\right)^3 =$

30) $\left(\frac{xy^5}{2xy^3}\right)^{-6} =$

Negative Exponents and Negative Bases

✎ **Simplify.**

1) $-4^{-2} =$

2) $-7^{-1} =$

3) $-5^{-2} =$

4) $-x^{-9} =$

5) $10x^{-2} =$

6) $-7x^{-4} =$

7) $-15x^{-4} =$

8) $-15x^{-7}y^{-4} =$

9) $32x^{-9}y^{-3} =$

10) $45a^{-7}b^{-3} =$

11) $-25x^3 y^{-5} =$

12) $-\dfrac{18}{x^{-9}} =$

13) $-\dfrac{13x}{a^{-8}} =$

14) $\left(-\dfrac{1}{3}\right)^{-4} =$

15) $\left(-\dfrac{3}{4}\right)^{-3} =$

16) $-\dfrac{12}{a^{-6}b^{-4}} =$

17) $-\dfrac{48x}{x^{-6}} =$

18) $-\dfrac{a^{-12}}{b^{-5}} =$

19) $-\dfrac{27}{x^{-5}} =$

20) $\dfrac{12b}{-48c^{-6}} =$

21) $\dfrac{24ab}{a^{-4}b^{-3}} =$

22) $-\dfrac{8n^{-7}}{40p^{-9}} =$

23) $\dfrac{9ab^{-6}}{-5c^{-2}} =$

24) $\left(\dfrac{2a}{3c}\right)^{-4} =$

25) $\left(-\dfrac{8x}{5yz}\right)^{-2} =$

26) $\dfrac{9ab^{-6}}{-4c^{-3}} =$

27) $\left(-\dfrac{x^3}{x^4}\right)^{-5} =$

28) $\left(-\dfrac{x^{-3}}{3x^3}\right)^{-3} =$

29) $\left(-\dfrac{x^{-6}}{x^4}\right)^{-3} =$

Answers of Worksheets – Chapter 1

Adding and Subtracting Integers

1) −20
2) −57
3) −8
4) −7
5) 31
6) 37
7) 12
8) −27
9) 6
10) 17
11) −25
12) 52
13) −116
14) 44
15) 11
16) 55
17) 100
18) 45
19) 58
20) 176
21) 126
22) 64
23) 84
24) 36
25) 133
26) 61

Multiplying and Dividing Integers

1) 63
2) −30
3) −150
4) 225
5) 1,092
6) −121
7) 500
8) −1,045
9) −420
10) −5,400
11) −17
12) 6
13) 11
14) −3
15) −57
16) 13
17) −18
18) 26
19) −17
20) −21
21) 23
22) 19
23) 30
24) −15
25) 26
26) −9

Order of Operations

1) 47
2) −38
3) 97
4) −32
5) 41
6) 27
7) −110
8) 115
9) 152
10) 40
11) 59
12) 62
13) −32
14) 69
15) 1,050
16) 124
17) 14
18) 83
19) 155
20) 32

Integers and Absolute Value

1) 19
2) 32
3) 50
4) 31
5) 57
6) 76
7) 42
8) 101
9) 28
10) 49
11) 13
12) 78
13) 100
14) 0
15) 105
16) 77
17) 88
18) 29
19) 112
20) 120
21) 0
22) -12
23) 72
24) 48

25) 55   26) 24   27) 156   28) 418

Sets

1) {1, 2, 3, 4, 5, 7, 8, 12}
2) {1, 2, 3, 5, 7, 8, 9, 11, 12}
3) {2, 4, 5, 7, 9, 11}
4) {2}
5) { } (empty set)
6) {5, 7}
7) {1, 2, 3, 4, 5, 7, 8, 9, 11, 12}
8) {5, 7}
9) { } (empty set)
10) {2}
11) {2, 3, 5, 6, 7, 8}
12) {2, 4, 5, 6, 7, 8}
13) {3, 4, 5, 6, 7, 8}
14) {7, 8}
15) {5, 7}
16) {6, 7}
17) {2, 3, 4, 5, 6, 7, 8}
18) {5, 6, 7}
19) {7}
20) {5, 7, 8}

Scientific Notation

1) 2.26×10^{-1}
2) 5×10^{-2}
3) 4.8×10^0
4) 9×10^1
5) 1.2×10^2
6) 1.23×10^{-1}
7) 8.2×10^1
8) 5.4×10^3
9) 2.46×10^3
10) 7.53×10^4
11) 61×10^6
12) 9×10^{-5}
13) 4.68×10^5
14) 4.58×10^{-3}
15) 8.7×10^{-5}
16) 3.18×10^7
17) 9.5×10^5
18) 9×10^9
19) 7×10^{-4}
20) 4.1×10^{-4}
21) 0.04
22) 0.0007
23) 4,300,000
24) 0.0007
25) 0.0087
26) 1,200,000
27) 35,000
28) 189,000
29) 0.000013
30) 0.00073

Powers of Products and Quotients

1) 4^6
2) 5^4
3) 3^9
4) 24^2
5) 15^{20}
6) 7^{20}
7) 9^6
8) 4^{18}
9) $343x^{21}$
10) $64x^8y^6$
11) $81x^{12}y^8$
12) $16x^4y^4$
13) $27x^{15}y^6$
14) $64x^9y^6$
15) $32x^{20}$
16) $36x^{12}$
17) $49x^{24}y^{10}$
18) $125x^{33}$
19) $2,304x^6$
20) $729x^{42}y^9$
21) $625x^{16}y^9$
22) $243x^{15}y^8$
23) $256x^2y^6$
24) $\frac{512}{x^6}$
25) x^7
26) $\frac{36y^2}{x^8}$
27) $\frac{1}{x^3y^6}$
28) x^6y^{18}
29) $\frac{125y^{21}}{x^3}$

30) $\frac{64}{y^{12}}$

Negative Exponents and Negative Bases

1) $-\frac{1}{16}$
2) $-\frac{1}{7}$
3) $-\frac{1}{25}$
4) $-\frac{1}{x^9}$
5) $\frac{10}{x^2}$
6) $-\frac{7}{x^4}$
7) $-\frac{15}{x^4}$
8) $-\frac{15}{x^7 y^4}$
9) $\frac{32}{x^9 y^3}$
10) $\frac{45}{a^7 b^3}$

11) $-\frac{25x^3}{y^5}$
12) $-18x^9$
13) $-13xa^8$
14) 81
15) $-\frac{64}{27}$
16) $-12a^6 b^4$
17) $-48x^7$
18) $-\frac{b^5}{a^{12}}$
19) $-27x^5$
20) $-\frac{bc^6}{4}$
21) $24a^5 b^4$

22) $-\frac{p^9}{5n^7}$
23) $-\frac{9ac^2}{5b^6}$
24) $\frac{81c^4}{16a^4}$
25) $\frac{25y^2 z^2}{64x^2}$
26) $-\frac{9ac^3}{4b^6}$
27) $-x^5$
28) $-27x^{18}$
29) $-x^{30}$

Chapter 2:
Radicals Expressions

Topics that you will practice in this chapter:

- ✓ Square Roots
- ✓ Simplifying Radical Expressions
- ✓ Simplifying Radical Expressions Involving Fractions
- ✓ Multiplying Radical Expressions
- ✓ Adding and Subtracting Radical Expressions

Mathematics is no more computation than typing is literature.
— *John Allen Paulos*

Square Roots

✎ **Find the value each square root.**

1) $\sqrt{64} =$ ___

2) $\sqrt{4} =$ ___

3) $\sqrt{289} =$ ___

4) $\sqrt{0.25} =$ ___

5) $\sqrt{0.01} =$ ___

6) $\sqrt{0.09} =$ ___

7) $\sqrt{1,600} =$ ___

8) $\sqrt{2.25} =$ ___

9) $\sqrt{0} =$ ___

10) $\sqrt{0.04} =$ ___

11) $\sqrt{0.36} =$ ___

12) $\sqrt{0.81} =$ ___

13) $\sqrt{0.49} =$ ___

14) $\sqrt{1.21} =$ ___

15) $\sqrt{1.69} =$ ___

16) $\sqrt{0.16} =$ ___

17) $\sqrt{529} =$ ___

18) $\sqrt{625} =$ ___

19) $\sqrt{0.81} =$ ___

20) $\sqrt{20} =$ ___

21) $\sqrt{50} =$ ___

22) $\sqrt{676} =$ ___

23) $\sqrt{270} =$ ___

24) $\sqrt{32} =$ ___

✎ **Evaluate.**

25) $\sqrt{4} \times \sqrt{16} =$ _____

26) $\sqrt{49} \times \sqrt{64} =$ _____

27) $\sqrt{2} \times \sqrt{8} =$ _____

28) $\sqrt{17} \times \sqrt{17} =$ _____

29) $\sqrt{13} \times \sqrt{13} =$ _____

30) $\sqrt{15} \times \sqrt{15} =$ _____

31) $\sqrt{19} + \sqrt{19} =$ _____

32) $\sqrt{1} + \sqrt{1} =$ _____

33) $8\sqrt{7} - 2\sqrt{7} =$ _____

34) $7\sqrt{10} \times 6\sqrt{10} =$ _____

35) $9\sqrt{5} \times 2\sqrt{5} =$ _____

36) $8\sqrt{3} - \sqrt{12} =$ _____

Simplifying Radical Expressions

✎ **Simplify.**

1) $\sqrt{13y^2} =$

2) $\sqrt{60x^3} =$

3) $\sqrt[3]{27a} =$

4) $\sqrt{81x^2} =$

5) $\sqrt{150a} =$

6) $\sqrt[3]{135w^3} =$

7) $\sqrt{200x} =$

8) $\sqrt{192v} =$

9) $\sqrt[3]{64x} =$

10) $\sqrt{84x^3} =$

11) $\sqrt{121x^2} =$

12) $\sqrt[3]{48a} =$

13) $\sqrt{480} =$

14) $\sqrt{1,575p^2} =$

15) $\sqrt{108m^6} =$

16) $\sqrt{198x^3y^2} =$

17) $\sqrt{169x^2y^3} =$

18) $\sqrt{25a^6} =$

19) $\sqrt{50x^2y^3} =$

20) $\sqrt[3]{512y^3} =$

21) $2\sqrt{144x^2} =$

22) $3\sqrt{400x^2} =$

23) $\sqrt[3]{189xy^4} =$

24) $\sqrt[3]{1,331x^3y^5} =$

25) $3\sqrt{150a} =$

26) $\sqrt[3]{729y} =$

27) $3\sqrt{18xyr^3} =$

28) $6\sqrt{225x^2yz^6} =$

29) $3\sqrt[3]{125x^3y^2} =$

30) $7\sqrt{12a^2bc^4} =$

31) $4\sqrt[3]{1,000x^9y^{15}} =$

Multiplying Radical Expressions

✎ **Simplify.**

1) $\sqrt{11} \times \sqrt{11} =$

2) $\sqrt{5} \times \sqrt{15} =$

3) $\sqrt{3} \times \sqrt{12} =$

4) $\sqrt{20} \times \sqrt{25} =$

5) $\sqrt{5} \times (-2)\sqrt{35} =$

6) $2\sqrt{12} \times \sqrt{3} =$

7) $4\sqrt{24} \times \sqrt{6} =$

8) $\sqrt{5} \times (-\sqrt{75}) =$

9) $\sqrt{88} \times \sqrt{40} =$

10) $2\sqrt{45} \times 4\sqrt{105} =$

11) $\sqrt{32}(2 + \sqrt{2}) =$

12) $\sqrt{13x^2} \times \sqrt{13x} =$

13) $-7\sqrt{12} \times \sqrt{3} =$

14) $5\sqrt{19x^3} \times \sqrt{19x^3} =$

15) $\sqrt{15x^2} \times \sqrt{5x} =$

16) $-8\sqrt{2x} \times \sqrt{6x^5} =$

17) $-4\sqrt{5x} \times 5\sqrt{45x^2} =$

18) $-3\sqrt{27}(3 + \sqrt{15}) =$

19) $\sqrt{8x}\,(3 - \sqrt{2x}) =$

20) $\sqrt{5x}(10\sqrt{5x} + \sqrt{40}) =$

21) $\sqrt{18r}\,(6 + \sqrt{6}) =$

22) $-12\sqrt{3x} \times 3\sqrt{15x^3} =$

23) $-4\sqrt{27x} \times 6\sqrt{3x}$

24) $-3\sqrt{10v^2}\,(-3\sqrt{15v}) =$

25) $(\sqrt{8} - 3)(\sqrt{8} + \sqrt{9}) =$

26) $(-3\sqrt{5} + 7)(\sqrt{5} - 3) =$

27) $(3 - 4\sqrt{5})(-2 + \sqrt{4}) =$

28) $(13 - 2\sqrt{5})(3 - \sqrt{5}) =$

29) $(5 - \sqrt{3x})(5 + \sqrt{3x}) =$

30) $(-6 + 3\sqrt{3r})(-6 + \sqrt{3r}) =$

31) $(-\sqrt{5n} + 8)(-\sqrt{5} - 8) =$

32) $(-3 + 3\sqrt{2})(3 - 2\sqrt{2x}) =$

Simplifying Radical Expressions Involving Fractions

✎ **Simplify.**

1) $\frac{\sqrt{3}}{\sqrt{2}} =$

2) $\frac{\sqrt{24}}{\sqrt{40}} =$

3) $\frac{\sqrt{12}}{2\sqrt{6}} =$

4) $\frac{21}{\sqrt{5}} =$

5) $\frac{15\sqrt{8r}}{\sqrt{m^5}} =$

6) $\frac{8\sqrt{2}}{\sqrt{m}} =$

7) $\frac{15\sqrt{25n^2}}{5\sqrt{15n}} =$

8) $\frac{\sqrt{8x^3y^5}}{\sqrt{2y^2x^4}} =$

9) $\frac{2}{2+\sqrt{5}} =$

10) $\frac{2-12\sqrt{x}}{\sqrt{24x}} =$

11) $\frac{2\sqrt{x}}{\sqrt{x}-\sqrt{y}} =$

12) $\frac{3-\sqrt{5}}{5-\sqrt{3}} =$

13) $\frac{5+\sqrt{12}}{5-\sqrt{3}} =$

14) $\frac{8}{-3-3\sqrt{3}} =$

15) $\frac{5}{2+\sqrt{15}} =$

16) $\frac{\sqrt{11}-\sqrt{7}}{\sqrt{7}-\sqrt{11}} =$

17) $\frac{\sqrt{8}+\sqrt{2}}{\sqrt{2}-\sqrt{8}} =$

18) $\frac{2\sqrt{2}-\sqrt{3}}{3\sqrt{2}+\sqrt{3}} =$

19) $\frac{\sqrt{11}+3\sqrt{5}}{3-\sqrt{11}} =$

20) $\frac{\sqrt{7}+\sqrt{13}}{13-\sqrt{7}} =$

21) $\frac{\sqrt{125a^7b^5}}{\sqrt{5ab^4}} =$

22) $\frac{72\sqrt{24m^3}}{9\sqrt{m}} =$

Adding and Subtracting Radical Expressions

✎ **Simplify.**

1) $\sqrt{5} + \sqrt{20} =$

2) $7\sqrt{44} + 7\sqrt{11} =$

3) $9\sqrt{3} - 3\sqrt{12} =$

4) $8\sqrt{9} - 5\sqrt{3} =$

5) $9\sqrt{80} - 9\sqrt{20} =$

6) $-\sqrt{32} - 5\sqrt{8} =$

7) $-12\sqrt{16} - 9\sqrt{64} =$

8) $15\sqrt{8} + 6\sqrt{32} =$

9) $16\sqrt{9} - 12\sqrt{36} =$

10) $-7\sqrt{7} + 11\sqrt{63} =$

11) $-24\sqrt{13} + 18\sqrt{117} =$

12) $25\sqrt{5} - 12\sqrt{45} =$

13) $-8\sqrt{99} + 2\sqrt{11} =$

14) $6\sqrt{3} - 2\sqrt{12} =$

15) $8\sqrt{20} + 3\sqrt{5} =$

16) $5\sqrt{28} - 8\sqrt{63} =$

17) $\sqrt{144} - \sqrt{121} =$

18) $6\sqrt{18} - 2\sqrt{2} =$

19) $-12\sqrt{7} + 21\sqrt{28} =$

20) $5\sqrt{60} - 5\sqrt{15} =$

21) $6\sqrt{54} - 3\sqrt{6} =$

22) $-4\sqrt{3} + 8\sqrt{75} =$

23) $-9\sqrt{20} - 7\sqrt{5} =$

24) $-\sqrt{216x} + 6\sqrt{6x} =$

25) $\sqrt{14y^2} + y\sqrt{126} =$

26) $\sqrt{11mn^2} + n\sqrt{99m} =$

27) $-8\sqrt{48a} - 2\sqrt{3a} =$

28) $-15\sqrt{17ab} - 10\sqrt{68ab} =$

29) $\sqrt{92x^2y} + x\sqrt{23y} =$

30) $5\sqrt{5a} + 4\sqrt{80a} =$

Answers of Worksheets – Chapter 2

Square Roots

1) 8
2) 2
3) 17
4) 0.5
5) 0.1
6) 0.3
7) 40
8) 1.5
9) 0
10) 0.2
11) 0.6
12) 0.9
13) 0.7
14) 1.1
15) 1.3
16) 0.4
17) 23
18) 25
19) 0.9
20) $2\sqrt{5}$
21) $5\sqrt{2}$
22) 26
23) $3\sqrt{30}$
24) $4\sqrt{2}$
25) 8
26) 56
27) 4
28) 17
29) 13
30) 15
31) $2\sqrt{19}$
32) 2
33) $6\sqrt{7}$
34) 420
35) 90
36) $6\sqrt{3}$

Simplifying radical expressions

1) $y\sqrt{13}$
2) $2x\sqrt{15x}$
3) $3\sqrt[3]{a}$
4) $9x$
5) $5\sqrt{6a}$
6) $3w\sqrt[3]{5}$
7) $10\sqrt{2x}$
8) $8\sqrt{3v}$
9) $4\sqrt[3]{x}$
10) $2x\sqrt{21x}$
11) $11x$
12) $2\sqrt[3]{6a}$
13) $4\sqrt{30}$
14) $15p\sqrt{7}$
15) $6m^3\sqrt{3}$
16) $3x.y\sqrt{22x}$
17) $13xy\sqrt{y}$
18) $5a^3$
19) $5xy\sqrt{2y}$
20) $8y$
21) $24x$
22) $60x$
23) $3y\sqrt[3]{7xy}$
24) $11xy\sqrt[3]{y^2}$
25) $15\sqrt{6a}$
26) $9\sqrt[3]{y}$
27) $9r\sqrt{2xyr}$
28) $90xz^3\sqrt{y}$
29) $15x\sqrt[3]{y^2}$
30) $14ac^2\sqrt{b}$
31) $40x^3y^{15}$

Multiplying radical expressions

1) 11
2) $5\sqrt{3}$
3) 6
4) $10\sqrt{5}$
5) $-10\sqrt{7}$
6) 12
7) 48
8) $-5\sqrt{15}$
9) $8\sqrt{55}$
10) $120\sqrt{21}$
11) $8\sqrt{2}+8$
12) $13x\sqrt{x}$

13) -42

14) $95x^3$

15) $5x\sqrt{3x}$

16) $-16x^3\sqrt{3}$

17) $-300x\sqrt{x}$

18) $-27\sqrt{3} - 27\sqrt{5}$

19) $6\sqrt{2x} - 4x$

20) $50x + 2\sqrt{50x}$

21) $18\sqrt{2r} + 6\sqrt{3r}$

22) $-108x^2\sqrt{5}$

23) $-216x$

24) $47v\sqrt{6v}$

25) -1

26) $16\sqrt{5} - 36$

27) 0

28) $49 - 19\sqrt{5}$

29) $28 - 3x$

30) $9r - 24\sqrt{3r} + 36$

31) $5\sqrt{n} - 64$

32) $-9 + 6\sqrt{2x} + 9\sqrt{2} - 12\sqrt{x}$

Simplifying radical expressions involving fractions

1) $\frac{\sqrt{6}}{2}$

2) $\frac{\sqrt{15}}{5}$

3) $\frac{\sqrt{2}}{2}$

4) $\frac{21\sqrt{5}}{5}$

5) $\frac{30\sqrt{2mr}}{m^3}$

6) $\frac{8\sqrt{2m}}{m}$

7) $\sqrt{15n}$

8) $\frac{2y\sqrt{xy}}{x}$

9) $-2(2 - \sqrt{2})$

10) $\frac{\sqrt{6x}(1-6\sqrt{x})}{6x}$

11) $\frac{2\sqrt{x}(\sqrt{x}+\sqrt{y})}{x-y}$

12) $\frac{15+ 3\sqrt{3}- 5\sqrt{5}-\sqrt{15}}{22}$

13) $\frac{31+15\sqrt{3}}{22}$

14) $-\frac{4(\sqrt{3}-1)}{3}$

15) $-\frac{5(2-\sqrt{15})}{11}$

16) -1

17) -3

18) $\frac{3-\sqrt{6}}{3}$

19) $-\frac{3\sqrt{11}+11+9\sqrt{5}+3\sqrt{55}}{2}$

20) $\frac{13\sqrt{7}+7+13\sqrt{13}+\sqrt{91}}{162}$

21) $5a^3\sqrt{b}$

22) $16\sqrt{6}\, m$

Adding and subtracting radical expressions

1) $2\sqrt{5}$

2) $21\sqrt{11}$

3) $-5\sqrt{3}$

4) $24 - 5\sqrt{3}$

5) $18\sqrt{5}$

6) $-14\sqrt{2}$

7) -120

8) $54\sqrt{2}$

9) -24

10) $26\sqrt{7}$

11) $30\sqrt{13}$

12) $-11\sqrt{5}$

13) $-22\sqrt{11}$

14) $2\sqrt{3}$

15) $19\sqrt{5}$

16) $-14\sqrt{7}$

17) 1

18) $16\sqrt{2}$

19) $30\sqrt{7}$

20) $5\sqrt{15}$

21) $15\sqrt{6}$

22) $36\sqrt{3}$

23) $-25\sqrt{5}$

24) 0

25) $4y\sqrt{14}$

26) $4n\sqrt{11m}$

27) $-34\sqrt{3a}$

28) $-35\sqrt{17ab}$

29) $-x\sqrt{23y}$

30) $21\sqrt{5a}$

Chapter 3:
Equations and Inequalities

Topics that you will practice in this chapter:

- ✓ One–Step Equations
- ✓ Multi–Step Equations
- ✓ Graphing Single–Variable Inequalities
- ✓ One–Step Inequalities
- ✓ Multi-Step Inequalities
- ✓ The Distributive Property
- ✓ Systems of Equations
- ✓ Systems of Equations Word Problems
- ✓ Systems of 3 variable Equations

"Life is a math equation. In order to gain the most, you have to know how to convert negatives into positives." − Anonymous

One–Step Equations

✍ **Find the answer for each equation.**

1) $3x = 90, x =$ ___

2) $5x = 35, x =$ ___

3) $9x = 36, x =$ ___

4) $25x = 150, x =$ ___

5) $x + 18 = 23, x =$ ___

6) $x - 3 = 8, x =$ ___

7) $x - 7 = 4, x =$ ___

8) $x + 22 = 30, x =$ ___

9) $x - 11 = 6, x =$ ___

10) $24 = 28 + x, x =$ ___

11) $x - 5 = 7, x =$ ___

12) $9 - x = -7, x =$ ___

13) $43 = -8 + x, x =$ ___

14) $x - 23 = -38, x =$ ___

15) $x + 45 = -27, x =$ ___

16) $42 = 56 - x, x =$ ___

17) $-18 + x = -32, x =$ ___

18) $x - 13 = 7, x =$ ___

19) $35 = x - 10, x =$ ___

20) $x - 8 = -21, x =$ ___

21) $x - 54 = -20, x =$ ___

22) $x - 42 = -47, x =$ ___

23) $x - 8 = 29, x =$ ___

24) $-93 = x - 51, x =$ ___

25) $x + 15 = 37, x =$ ___

26) $108 = 12x, x =$ ___

27) $x - 33 = 27, x =$ ___

28) $x - 12 = 23, x =$ ___

29) $72 - x = 18, x =$ ___

30) $x + 34 = 58, x =$ ___

31) $21 - x = -9, x =$ ___

32) $x - 59 = -80, x =$ ___

Multi–Step Equations

✏️ **Find the answer for each equation.**

1) $3x + 1 = 7$

2) $-x + 10 = 9$

3) $5x - 13 = 7$

4) $-(4 - x) = 5$

5) $3x - 8 = 16$

6) $15x - 13 = 17$

7) $3x - 28 = 2$

8) $9x + 21 = 39$

9) $14x + 17 = 45$

10) $-14(8 + x) = 70$

11) $8(10 + x) = 32$

12) $16 = -(x - 8)$

13) $5(7 - 3x) = 50$

14) $-19 = -(3x + 7)$

15) $30(3 + x) = 60$

16) $9(x - 12) = 54$

17) $-24 = 3x + 5x$

18) $5x + 28 = -2x - 7$

19) $9(5 + 4x) = -99$

20) $18 - x = -12 - 6x$

21) $4 - 4x = 28 - 2x$

22) $15 + 12x = -15 + 8x$

23) $54 = (-3x) - 8 + 8$

24) $12 = 7x - 18 + 5x$

25) $-18 = -9x - 42 + 5x$

26) $11x - 6 = -33 + 8x$

27) $8x - 42 = 3x + 3$

28) $-15 - 8x = 4(5 - x)$

29) $x - 9 = -5(9 - 2x)$

30) $14x - 65 = -x - 110$

31) $3x - 129 = -3(11 + 7x)$

32) $-7x - 20 = 2x + 43$

Graphing Single–Variable Inequalities

✎ **Draw a graph for each inequality.**

1) $x \leq 7$

2) $x \leq -1.5$

3) $x < -4$

4) $x > 2.5$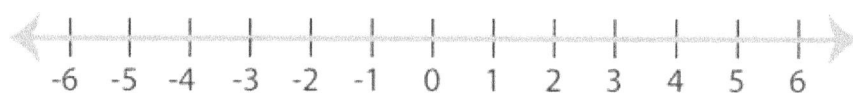

5) $x > 1.3$

6) $x < 4$

7) $x < 2.4$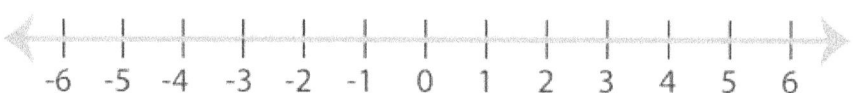

8) $x > -\frac{18}{10}$

One–Step Inequalities

✏ **Find the answer for each inequality and graph it.**

1) $x + 3 > -5$

2) $x - 4 < 1$

3) $7x < 42$

4) $13 + x > 12$

5) $x + 20 < 13$

6) $14x \leq 42$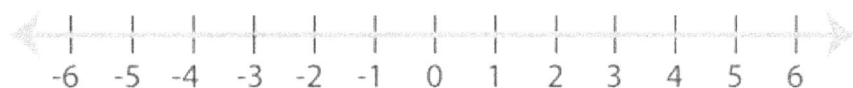

7) $11x \leq -44$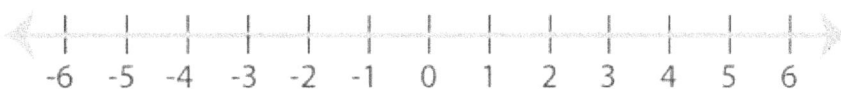

8) $x + 26 > 35$

Multi-Step Inequalities

✏️ **Calculate each inequality.**

1) $x - 8 \leq 12$

2) $9 - 3x \leq 18$

3) $4x - 7 \leq 9$

4) $8x - 9 \geq 15$

5) $x - 19 \geq 24$

6) $5x - 15 \leq 40$

7) $7x - 4 \leq 24$

8) $-18 + 8x \leq 22$

9) $9(x - 8) \leq 27$

10) $4x - 8 \leq 16$

11) $11x - 42 < 22$

12) $10x - 18 < 52$

13) $17 - 9x \geq -46$

14) $32 + 2x < 68$

15) $8 + 8x \geq 80$

16) $11 + 6x < 65$

17) $9x - 13 < 23$

18) $8(12 - 4x) \geq -68$

19) $-(2 + 5x) < 42$

20) $14 - 9x \geq -31$

21) $-5(x - 3) > 65$

22) $\dfrac{2x + 8}{3} \leq 12$

23) $\dfrac{8x + 16}{4} \leq 24$

24) $\dfrac{2x - 22}{9} > 8$

25) $7 + \dfrac{x}{4} < 21$

26) $\dfrac{32x}{16} - 4 < 6$

27) $\dfrac{12x + 36}{22} > 3$

28) $42 + \dfrac{x}{3} < 15$

The Distributive Property

✏ **Use the distributive property to simply each expression.**

1) $4(2 + 5x) =$

2) $5(2 + 4x) =$

3) $6(5x - 5) =$

4) $(6x - 3)(-7) =$

5) $(-4)(x + 8) =$

6) $(4 + 4x)6 =$

7) $(-5)(8 - 7x) =$

8) $-(-3 - 12x) =$

9) $(-8x + 3)(-5) =$

10) $(-5)(x - 11) =$

11) $-(8 - 2x) =$

12) $3(7 + 4x) =$

13) $4(8 + 3x) =$

14) $(-8x + 2)5 =$

15) $(4 - 7x)(-9) =$

16) $(-12)(3x + 5) =$

17) $(9 - 3x)5 =$

18) $4(4 + 7x) =$

19) $12(3x - 6) =$

20) $(-7x + 5)4 =$

21) $(4 - 9x)(-2) =$

22) $(-15)(2x - 3) =$

23) $(14 - 3x)3 =$

24) $(-5)(10x - 4) =$

25) $(5 - 7x)(-12) =$

26) $(-8)(2x + 9) =$

27) $(-5 + 8x)(-7) =$

28) $(-6)(2 - 15x) =$

29) $13(4x - 6) =$

30) $(-15x + 13)(-4) =$

31) $(-9)(3x - 2) + 2(x + 5) =$

32) $(-9)(2x + 2) - (7 + 4x) =$

Systems of Equations

✎ **Calculate each system of equations.**

1) $-6x + 7y = 8$ $x = $ ___
   $x + 4y = 9$ $y = $ ___

2) $-4x + 12y = 12$ $x = $ ___
   $14x - 16y = 10$ $y = $ ___

3) $y = -9$ $x = $ ___
   $2x - 5y = 12$ $y = $ ___

4) $4y = -4x + 20$ $x = $ ___
   $8x - 2y = -12$ $y = $ ___

5) $10x - 9y = -13$ $x = $ ___
   $-5x + 3y = 11$ $y = $ ___

6) $-6x - 8y = 10$ $x = $ ___
   $4x - 8y = 20$ $y = $ ___

7) $5x - 14y = -23$ $x = $ ___
   $-6x + 7y = 8$ $y = $ ___

8) $-4x + 3y = 3$ $x = $ ___
   $-x + 2y = 5$ $y = $ ___

9) $-4x + 5y = 15$ $x = $ ___
   $-3x + 4y = -10$ $y = $ ___

10) $-6x - 6y = -21$ $x = $ ___
    $-6x + 6y = -66$ $y = $ ___

11) $12x - 21y = 6$ $x = $ ___
    $-6x - 3y = -12$ $y = $ ___

12) $-4x - 4y = -14$ $x = $ ___
    $4x - 4y = 44$ $y = $ ___

13) $4x + 5y = 3$ $x = $ ___
    $3x - y = 6$ $y = $ ___

14) $3x - 2y = 2$ $x = $ ___
    $10x - 10y = 20$ $y = $ ___

15) $5x + 8y = 14$ $x = $ ___
    $-3x - 2y = -3$ $y = $ ___

16) $8x + 5y = 4$ $x = $ ___
    $-3x - 4y = 15$ $y = $ ___

Systems of Equations Word Problems

✎ **Find the answer for each word problem.**

1) Tickets to a movie cost $6 for adults and $4 for students. A group of friends purchased 9 tickets for $50.00. How many adults ticket did they buy? ____

2) At a store, Eva bought two shirts and five hats for $77.00. Nicole bought three same shirts and four same hats for $84.00. What is the price of each shirt? ____

3) A farmhouse shelters 10 animals, some are pigs, and some are ducks. Altogether there are 36 legs. How many pigs are there? ____

4) A class of 85 students went on a field trip. They took 24 vehicles, some cars and some buses. If each car holds 3 students and each bus hold 16 students, how many buses did they take? ____

5) A theater is selling tickets for a performance. Mr. Smith purchased 8 senior tickets and 10 child tickets for $248 for his friends and family. Mr. Jackson purchased 4 senior tickets and 6 child tickets for $132. What is the price of a senior ticket? $____

6) The difference of two numbers is 15. Their sum is 33. What is the bigger number? $____

7) The sum of the digits of a certain two–digit number is 7. Reversing its digits increase the number by 9. What is the number? ____

8) The difference of two numbers is 11. Their sum is 25. What are the numbers? _____

9) The length of a rectangle is 5 meters greater than 2 times the width. The perimeter of rectangle is 28 meters. What is the length of the rectangle? _____

10) Jim has 23 nickels and dimes totaling $2.40. How many nickels does he have? ____

Systems of 3 Variable Equations

✏️ **Solve each system of equations.**

1) $x = 3y - 3z + 8$ $x = \_\_$
   $z = 4x + 5y - 14$ $y = \_\_$
   $3y + 2z = 14$ $z = \_\_$

2) $6x - 6y = -12$ $x = \_\_$
   $2z = -6x - 6y + 18$ $y = \_\_$
   $-8x + 10y + 2z = 16$ $z = \_\_$

3) $4x - 8z = 40$ $x = \_\_$
   $-6x + 2y - 8z = 40$ $y = \_\_$
   $-8x + 4y + 6z = -30$ $z = \_\_$

4) $2x - 4y + 2z = -12$ $x = \_\_$
   $2x + 10z = -24$ $y = \_\_$
   $-2x + 12y + 8z = 6$ $z = \_\_$

5) $x - y - 2z = -6$ $x = \_\_$
   $3x + 2y = -25$ $y = \_\_$
   $-4x + y - z = 12$ $z = \_\_$

6) $6x - y + 3z = -9$ $x = \_\_$
   $5x + 5y - 5z = 20$ $y = \_\_$
   $3x - y + 4z = -5$ $z = \_\_$

7) $-5x + 3y + 6z = 4$ $x = \_\_$
   $-3x + y + 5z = -5$ $y = \_\_$
   $-4x + 2y + z = 13$ $z = \_\_$

8) $-6x + 5y + 2z = -11$ $x = \_\_$
   $-2x + y + 4z = -9$ $y = \_\_$
   $4x - 5y + 5z = -4$ $z = \_\_$

9) $4x + 4y + z = 24$ $x = \_\_$
   $2x - 4y + z = 0$ $y = \_\_$
   $5x - 4y - 5z = 12$ $z = \_\_$

10) $-10x + 10y + 6z = -46$ $x = \_\_$
    $-10x + 6y - 6z = -22$ $y = \_\_$
    $-12x + 12z = -24$ $z = \_\_$

Answers of Worksheets – Chapter 3

One–Step Equations

1) 30	9) 17	17) −14	25) 22
2) 7	10) −4	18) 20	26) 9
3) 4	11) 12	19) 45	27) 60
4) 6	12) 16	20) −13	28) 35
5) 5	13) 51	21) 34	29) 54
6) 11	14) −15	22) −5	30) 24
7) 11	15) −72	23) 37	31) 30
8) 8	16) 14	24) −42	32) −21

Multi–Step Equations

1) 2	9) 2	17) −3	25) −6
2) 1	10) −13	18) −5	26) −9
3) 4	11) −6	19) −4	27) 9
4) 9	12) −8	20) −6	28) −8.75
5) 8	13) −1	21) −12	29) 4
6) 2	14) 4	22) −7.5	30) −3
7) 10	15) −1	23) −18	31) 4
8) 2	16) 18	24) 2.5	32) −7

Graphing Single–Variable Inequalities

1)

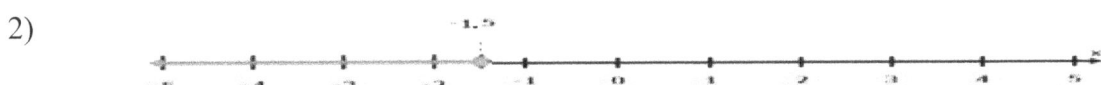

2)

3)

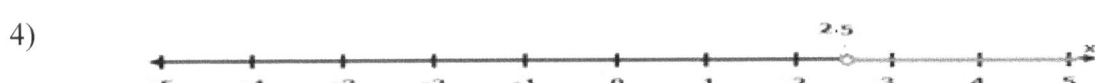

4)

5)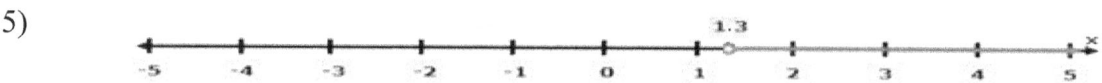

6)

7)

8)

One–Step Inequalities

1)

2)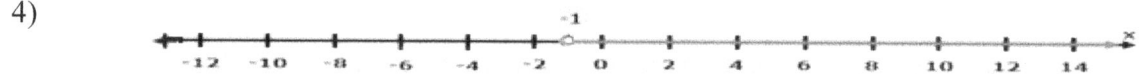

3)

4)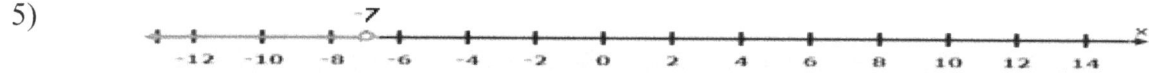

5)

6)

7)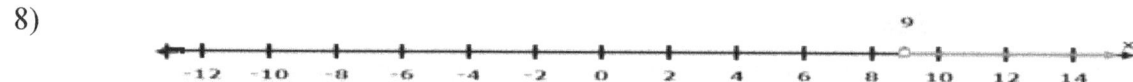

8)

Multi-Step Inequalities

1) $x \leq 20$ 5) $x \geq 43$ 9) $x \leq 11$ 13) $x \leq 7$

2) $x \geq -3$ 6) $x \leq 11$ 10) $x \leq 6$ 14) $x < 18$

3) $x \leq 4$ 7) $x \leq 4$ 11) $x < 64/11$ 15) $x \geq 9$

4) $x \geq 3$ 8) $x \leq 5$ 12) $x < 7$ 16) $x < 9$

17) $x < 4$
18) $x \leq 41/8$
19) $x > -44/5$
20) $x \leq 5$
21) $x < -10$
22) $x \leq 14$
23) $x \leq 10$
24) $x > 47$
25) $x < 56$
26) $x < 9/4$
27) $x > 2.5$
28) $x < -81$

The Distributive Property

1) $20x + 8$
2) $20x + 10$
3) $30x - 30$
4) $-42x + 21$
5) $-4x - 32$
6) $24x + 24$
7) $35x - 40$
8) $12x + 3$
9) $40x - 15$
10) $-5x + 55$
11) $2x - 8$
12) $12x + 21$
13) $12x + 32$
14) $-40x + 10$
15) $63x - 36$
16) $-36x - 60$
17) $-15x + 45$
18) $28x + 16$
19) $36x - 72$
20) $-28x + 20$
21) $18x - 8$
22) $-30x + 45$
23) $-9x + 42$
24) $-50x + 20$
25) $84x - 60$
26) $-16x - 72$
27) $56x - 35$
28) $90x - 12$
29) $52x - 78$
30) $60x - 52$
31) $-25x + 28$
32) $-22x - 25$

Systems of Equations

1) $x = 1, y = 2$
2) $x = 3, y = 2$
3) $x = -\frac{33}{2}$
4) $x = -\frac{1}{5}, y = \frac{26}{5}$
5) $x = -4, y = -3$
6) $x = 1, y = -2$
7) $x = 1, y = 2$
8) $x = \frac{9}{5}, y = \frac{17}{5}$
9) $x = -110, y = -85$
10) $x = -\frac{15}{4}, y = \frac{29}{4}$
11) $x = \frac{5}{3}, y = \frac{2}{3}$
12) $x = -\frac{15}{4}, y = \frac{29}{4}$
13) $x = \frac{33}{19}, y = -\frac{15}{19}$
14) $x = -2, y = -4$
15) $x = -\frac{2}{7}, y = \frac{27}{14}$
16) $x = \frac{91}{17}, y = -\frac{132}{17}$

Systems of Equations Word Problems

1) 7
2) $16
3) 8
4) 1
5) $21
6) 24
7) 43
8) 18, 7
9) 11 meters
10) 18

Systems of 3 variable equations

1) (2, 2, 4)
2) (1, 3, -3)
3) (0, 0, -5)
4) (3, 3, -3)
5) (-5, -5, 3)
6) (-1, 6, 1)
7) (-2, 4, -3)
8) (4, 3, -1)
9) (4, 2, 0)
10) (1, -3, -1)

Chapter 4:
Linear Functions

Topics that you will practice in this chapter:

- ✓ Relation and Function
- ✓ Finding Slope
- ✓ Graphing Lines Using Line Equation
- ✓ Writing Linear Equations
- ✓ Graphing Linear Inequalities
- ✓ Write an Equation from a Graph
- ✓ Finding Rate of Change, x–intercept and y–intercept
- ✓ Slope-Intercept Form
- ✓ Point-Slope Form
- ✓ Graphing Lines of Equations
- ✓ Equation of parallel or perpendicular lines
- ✓ Equations of horizontal and vertical lines
- ✓ Graphing Absolute Value Equation

"Sometimes the questions are complicated, and the answers are simple." – Dr. Seuss

Relation and Functions

State the domain and range of each relation. Then determine whether each relation is a function.

1)
Function:
....
Domain:
....
Range:
....

2)
Function:
....
Domain:
....
Range:
....

x	y
2	3
1	0
−1	−4
7	−4
9	5

3)
Function:
....
Domain:
....
Range:
....

4) $\{(6, -8), (3, -2), (2, 4), (3, 0), (5, 9)\}$
Function:
....
Domain:
....
Range:
....

5)
Function:
....
Domain:
....
Range:
....

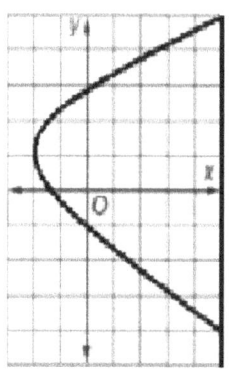

6)
Function:
....
Domain:
....
Range:
....

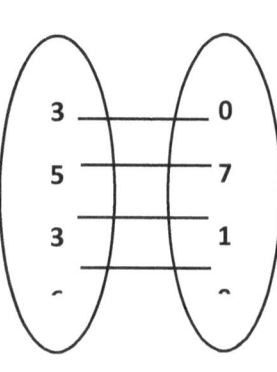

Finding Slope

✒ **Find the slope of each line.**

1) $y = 2x + 5$

2) $y = -x + 17$

3) $y = 4x + 16$

4) $y = -3x + 15$

5) $y = 27 + 7x$

6) $y = 11 - 4x$

7) $y = 7x + 14$

8) $y = -8x + 18$

9) $y = -9x + 15$

10) $y = 8x - 13$

11) $y = \frac{1}{5}x + 9$

12) $y = -\frac{3}{7}x + 19$

13) $-3x + 6y = 17$

14) $4x + 4y = 16$

15) $8y - 3x = 32$

16) $11y - 3x = 42$

✒ **Find the slope of the line through each pair of points.**

17) $(1, 8), (5, 16)$

18) $(-2, 14), (2, 18)$

19) $(7, -1), (3, 9)$

20) $(-4, -4), (2, 14)$

21) $(16, -1), (4, 11)$

22) $(-21, 5), (-10, 38)$

23) $(8, 11), (12, 19)$

24) $(22, -22), (10, 14)$

25) $(21, -15), (19, -13)$

26) $(11, 10), (7, -2)$

27) $(5, 4), (9, 16)$

28) $(34, -87), (22, 45)$

Graphing Lines Using Line Equation

✎ **Sketch the graph of each line.**

1) $y = x - 5$

2) $y = -3x + 4$

3) $x - 2y = 0$

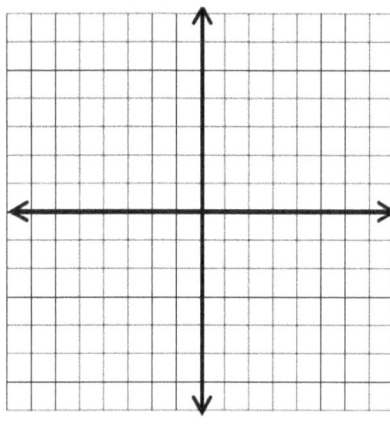 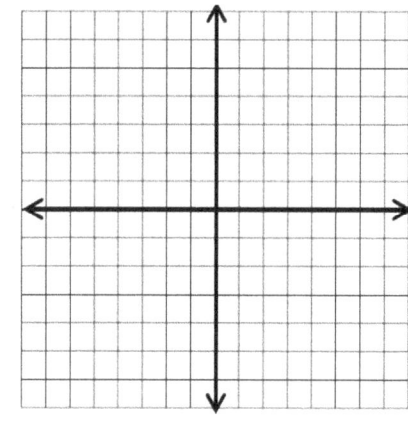 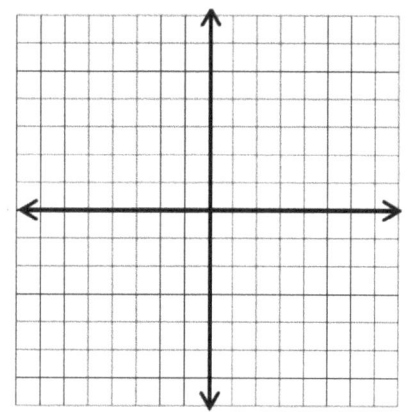

4) $x + y = -4$

5) $4x + 3y = -2$

6) $y - 3x + 6 = 0$

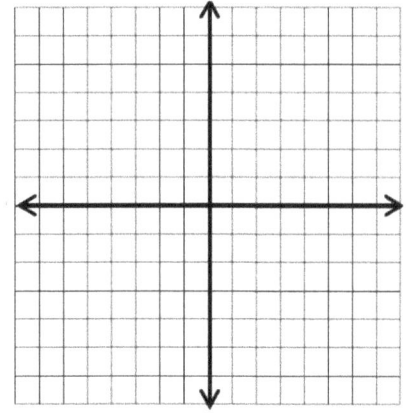 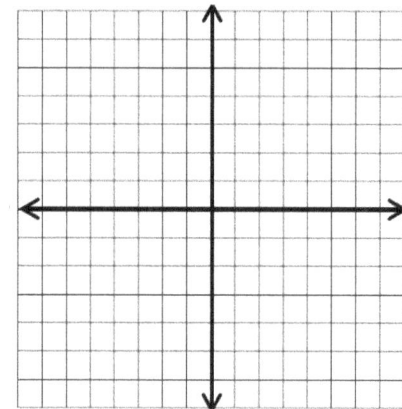 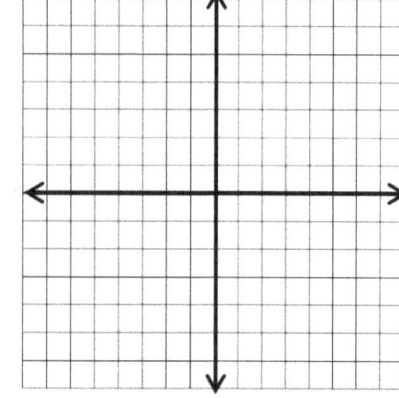

Writing Linear Equations

✎ **Write the equation of the line through the given points.**

1) Through: $(6, -10), (10, 14)$

2) Through: $(10, 4), (4, 22)$

3) Through: $(-6, 4), (2, 12)$

4) Through: $(15, 11), (3, -1)$

5) Through: $(-5, 33), (9, 5)$

6) Through: $(20, 5), (17, 2)$

7) Through: $(24, -4), (16, 4)$

8) Through: $(-18, 57), (33, -45)$

9) Through: $(10, 12), (8, 18)$

10) Through: $(25, 41), (33, -7)$

11) Through: $(-6, 9), (-8, -7)$

12) Through: $(8, 8), (4, -8)$

13) Through: $(6, -10), (10, 6)$

14) Through: $(10, -24), (-8, 12)$

15) Through: $(10, 10), (-2, -4)$

16) Through: $(-7, 35), (11, -31)$

✎ **Find the answer for each problem.**

17) What is the equation of a line with slope 3 and intercept 11? _____

18) What is the equation of a line with slope 5 and intercept 15? _____

19) What is the equation of a line with slope 7 and passes through point $(3, 2)$? _____

20) What is the equation of a line with slope -3 and passes through point $(-2, 5)$? _____

21) The slope of a line is -6 and it passes through point $(-2, 1)$. What is the equation of the line? _____

22) The slope of a line is 5 and it passes through point $(-4, 2)$. What is the equation of the line? _____

Graphing Linear Inequalities

✏️ **Sketch the graph of each linear inequality.**

1) $y > 3x - 5$

2) $y < 2x + 1$

3) $y \leq -4x - 5$

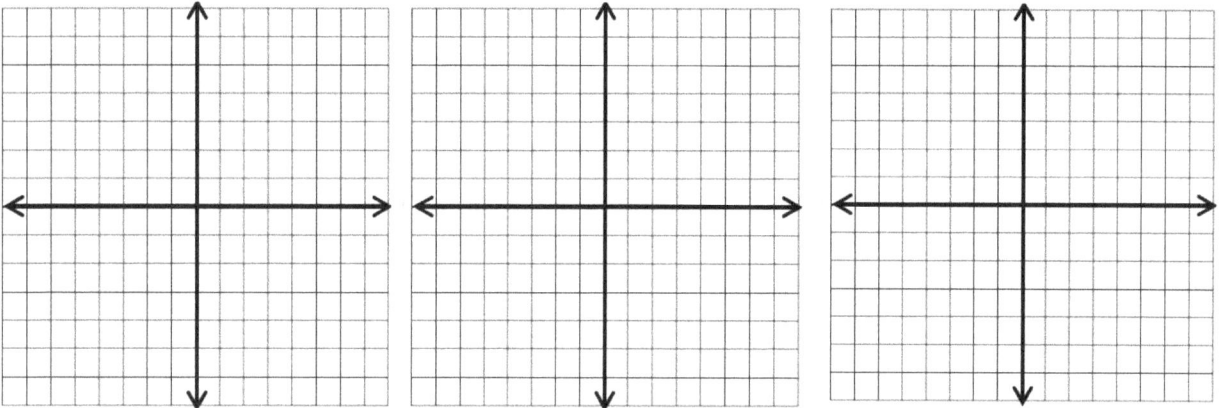

4) $2y \geq 12 + 4x$

5) $-5y < x - 15$

6) $3y \geq -9x + 6$

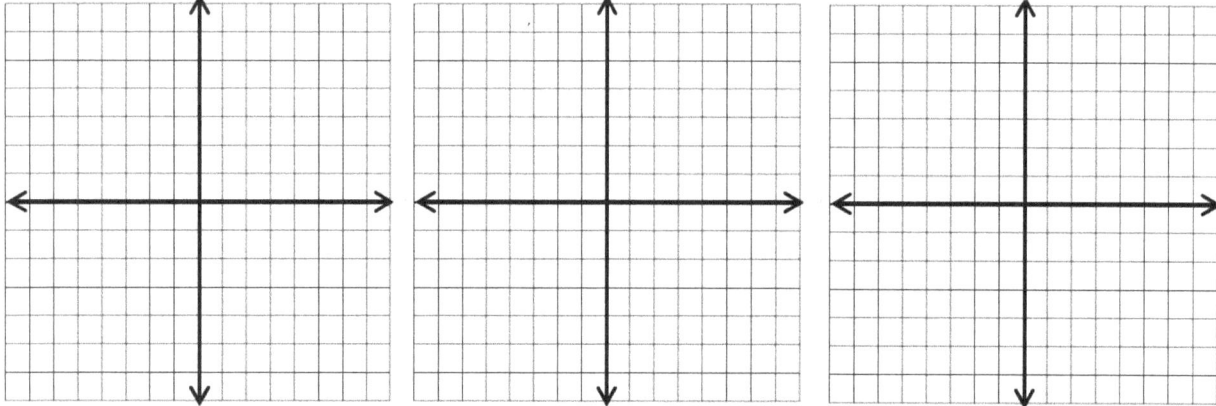

Write an Equation from a Graph

✍ Write the slope intercept form of the equation of each line

1)

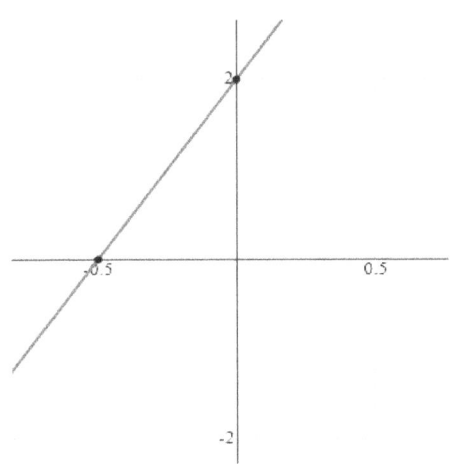

2)

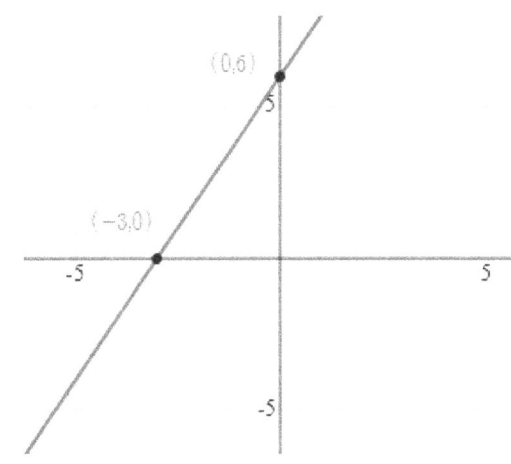

3)

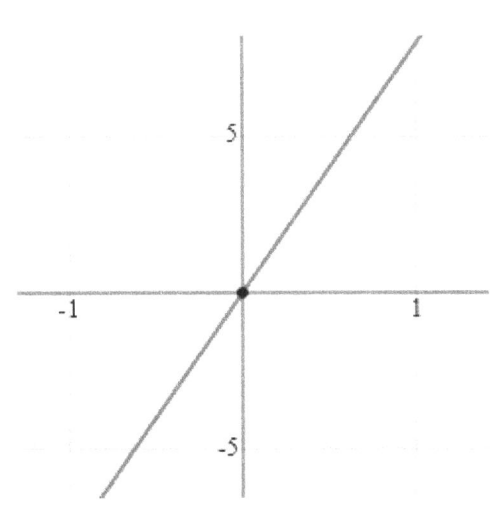

4)

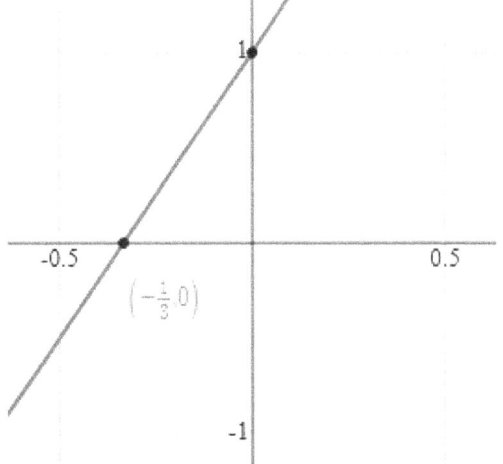

Rate of change

What is the average rate of change of the function?

1) $f(x) = 2x^2 + 3$ from $x = 2$ to $x = 5$?

2) $f(x) = -x^2 - 6$, from $x = 3$ to $x = 7$?

3) $f(x) = 2x^3 + 5$, from $x = 0$ to $x = 1$?

x and y intercepts

Find the x and y intercepts for the following equations.

1) $4x + 2y = 12$

2) $y = x + 4$

3) $3x = y + 18$

4) $x + y = -6$

5) $7x - 5y = 8$

6) $5y - 4x + 12 = 0$

7) $\frac{3}{5}x + \frac{1}{5}y + \frac{3}{4} = 0$

8) $6x - 24 = 0$

9) $28 - 7y = 0$

10) $-3x - 5y + 45 = 15$

Find the value of b: The line that passes through each pair of points has the given slope.

11) $(8, -3), (4, b), m = 2$

12) $(b, 5), (-5, 2), m = \frac{1}{2}$

13) $(-3, b), (3, 5), m = \frac{1}{3}$

14) $(-2, 2), (b, 9), m = 1\frac{3}{4}$

Slope–intercept Form

✏️ **Write the slope-intercept form of the equation of each line.**

1) $-15x + y = 7$

2) $-3(5x + y) = 36$

3) $-7x - 21y = -42$

4) $4x + 12 = -8y$

5) $2x - 5y = 15$

6) $14x - 10y = -20$

7) $27x - 9y = -54$

8) $6x - 5y + 36 = 0$

9) $-\frac{1}{4}y = -3x + 5$

10) $8 - 2y - 5x = 0$

11) $-2y = -3x - 8$

12) $12x + 7y = -21$

13) $4(x + 3y + 4) = 0$

14) $y - 6 = 2x + 5$

15) $4(y + 2) = 3(x - 2)$

16) $\frac{2}{5}y + \frac{3}{5}x + \frac{4}{5} = 0$

Point–slope Form

Find the slope of the following lines. Name a point on each line.

1) $y = 3(x + 5)$

2) $y + 2 = \frac{1}{4}(x - 3)$

3) $y + 1 = -2.5x$

4) $y - 4 = \frac{1}{3}(x - 4)$

5) $y + 5 = 0.6(x + 7)$

6) $y - 6 = -2x$

7) $y - 10 = -2(x - 9)$

8) $y + 18 = 0$

9) $y + 19 = 6(x + 1)$

10) $y - 14 = -3(x - 2)$

Write an equation in point–slope form for the line that passes through the given point with the slope provided.

11) $(9, -7), m = 5$

12) $(-2, 5), m = \frac{1}{2}$

13) $(0, -4), m = -3$

14) $(-a, b), m = n$

15) $(-8, 2), m = 4$

16) $(6, 1), m = -4$

17) $(-7, 12), m = \frac{1}{6}$

18) $(0, 13), m = 0$

19) $\left(-\frac{1}{2}, 2\right), m = \frac{1}{7}$

20) $(0, 0), m = -2$

Graphing Lines of Equations

🖊 **Sketch the graph of each line**

1) $y = 3x - 2$

2) $y = -\frac{1}{2}x + \frac{3}{2}$

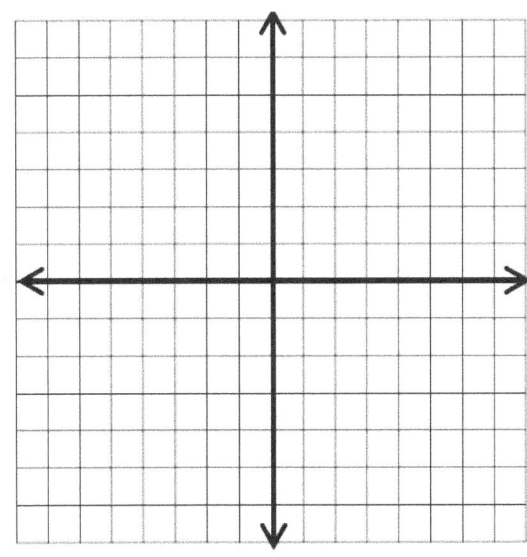
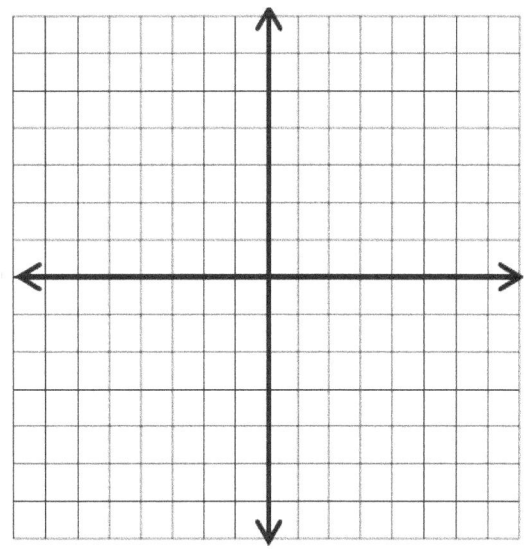

3) $4x - 5y = 12$

4) $-3x - y = 5$

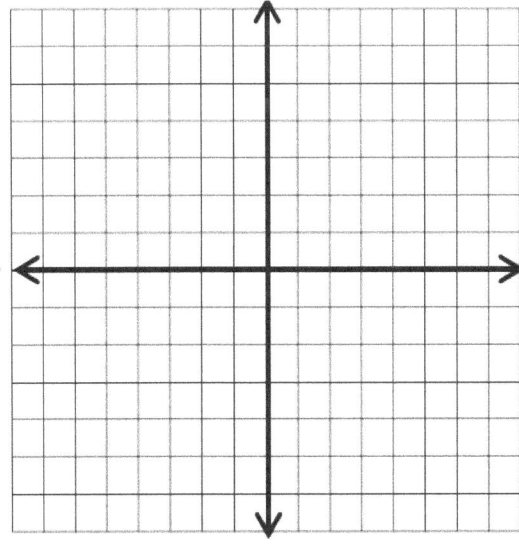
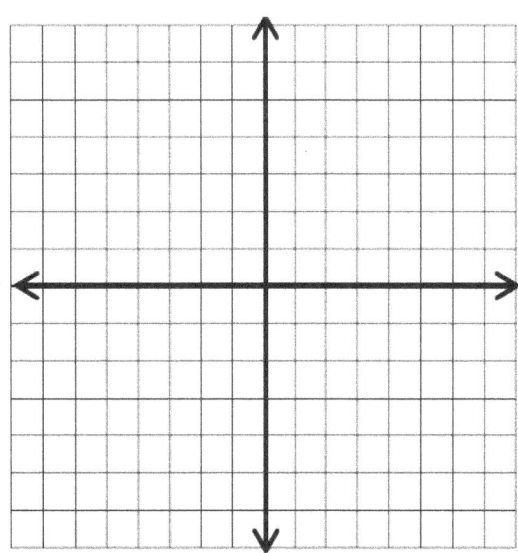

Equation of Parallel or Perpendicular Lines

✏ **Write an equation of the line that passes through the given point and is parallel to the given line.**

1) $(-3, -1), x + 2y = -10$

2) $(-3, 2), y = x - 4$

3) $(-3, 1), 4y = x - 7$

4) $(0, 1), -y + 2x - 8 = 0$

5) $(2, 8), y + 12 = 0$

6) $(1, 4), -4x - 2y = -5$

7) $(-3, 0), y = \frac{2}{3}x + 4$

8) $(-1, 3), -4x + y = -16$

9) $(1, -1), y = -\frac{1}{5}x - 2$

10) $(-3, -3), 2x + 10y = -20$

✏ **Write an equation of the line that passes through the given point and is perpendicular to the given line.**

11) $(-4, 0), 2x + y = -8$

12) $(-\frac{1}{2}, \frac{3}{4}), 9x - 6y = -9$

13) $(4, -8), y = -8$

14) $(9, -5), x = 9$

15) $(-8, 7), y = \frac{1}{4}x + 9$

16) $(\frac{1}{3}, \frac{2}{3}), y = -4x + 2$

17) $(-8, -4), y = \frac{7}{4}x + 10$

18) $(-8, 5), y = x + 13$

19) $(-4, -10), y = \frac{9}{4}x - 1$

20) $(5, 2), 5y - x + 8 = 13$

Equations of Horizontal and Vertical Lines

✎ Sketch the graph of each line.

1) $y = -2$

2) $y = 1$

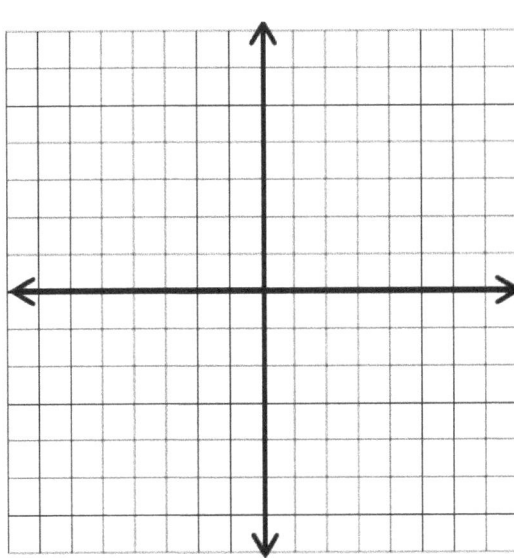
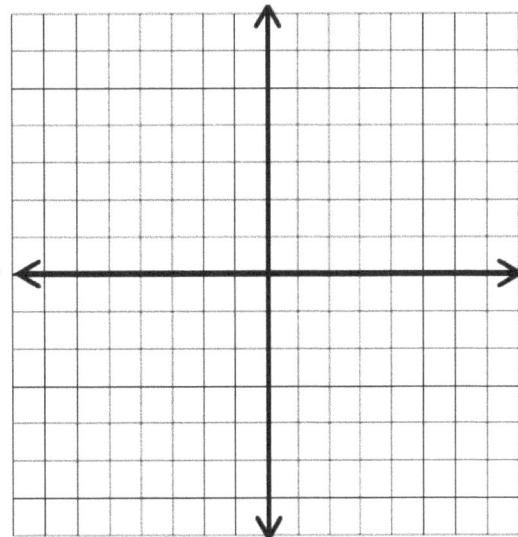

3) $x = -1$

4) $x = 2$

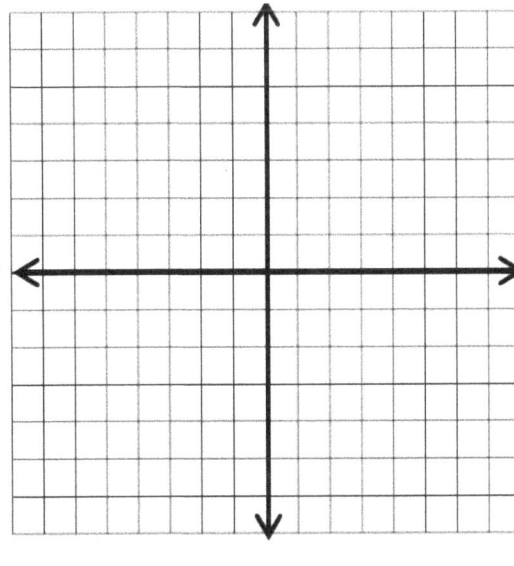
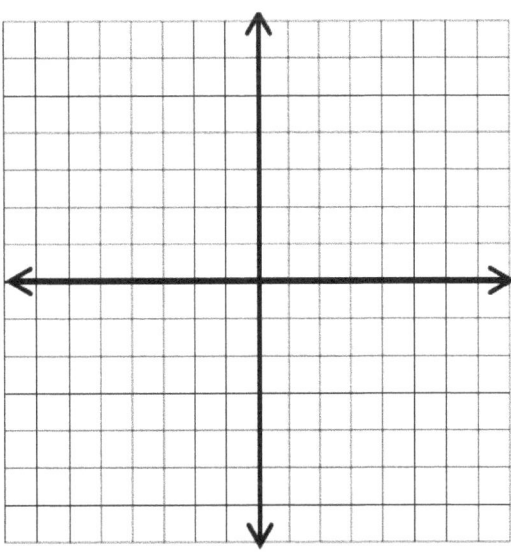

Graphing Absolute Value Equations

📝 **Graph each equation.**

1) $y = |x + 4|$ 2) $y = |x + 1|$ 3) $y = -|x| - 1$

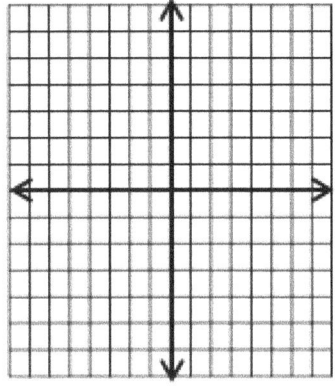   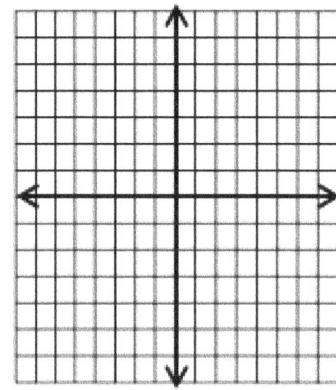   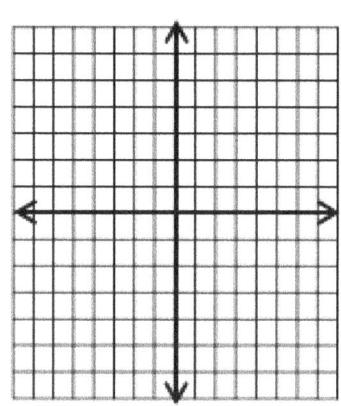

4) $y = |x - 2|$ 5) $y = -|x - 2|$ 6) $y = -2|2x + 2| + 4$

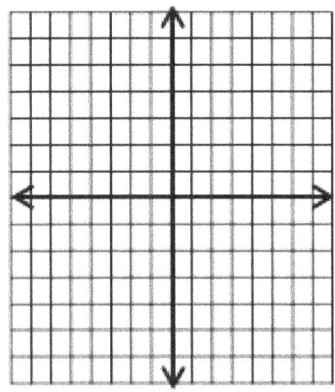   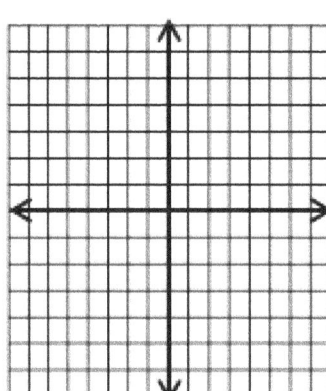   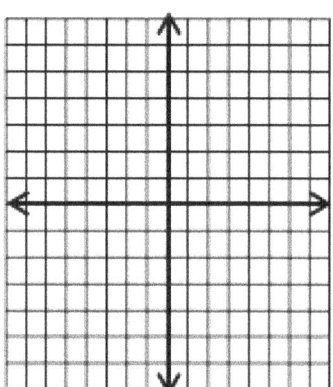

Answers of Worksheets – Chapter 4

Relation and Functions

1) No, $D_f = \{2, 4, 8, 9, 10\}$, $R_f = \{1, 0, 5, 7, 14\}$
2) Yes, $D_f = \{2, 1, -1, 7, 9\}$, $R_f = \{3, 0, -4, 5\}$
3) Yes, $D_f = (-\infty, \infty)$, $R_f = \{2, -\infty\}$
4) No, $D_f = \{6, 3, 2, 5\}$, $R_f = \{-8, -2, 4, 0, 9\}$
5) No, $D_f = [-2, \infty)$, $R_f = (-\infty, \infty)$
6) No, $D_f = \{3, 5, 6\}$, $R_f = \{0, 7, 1, 9\}$

Finding Slope

1) 2
2) -1
3) 4
4) -3
5) 7
6) -4
7) 7
8) -8
9) -9
10) 8
11) $\frac{1}{5}$
12) $-\frac{3}{7}$
13) $\frac{1}{2}$
14) -1
15) $\frac{3}{8}$
16) $\frac{3}{11}$
17) 2
18) 1
19) $-\frac{5}{2}$
20) 3
21) -1
22) 3
23) 2
24) -3
25) -1
26) 3
27) 3
28) -11

Graphing Lines Using Line Equation

1) $y = x - 5$ 2) $y = -3x + 4$ 3) $x - 2y = 0$

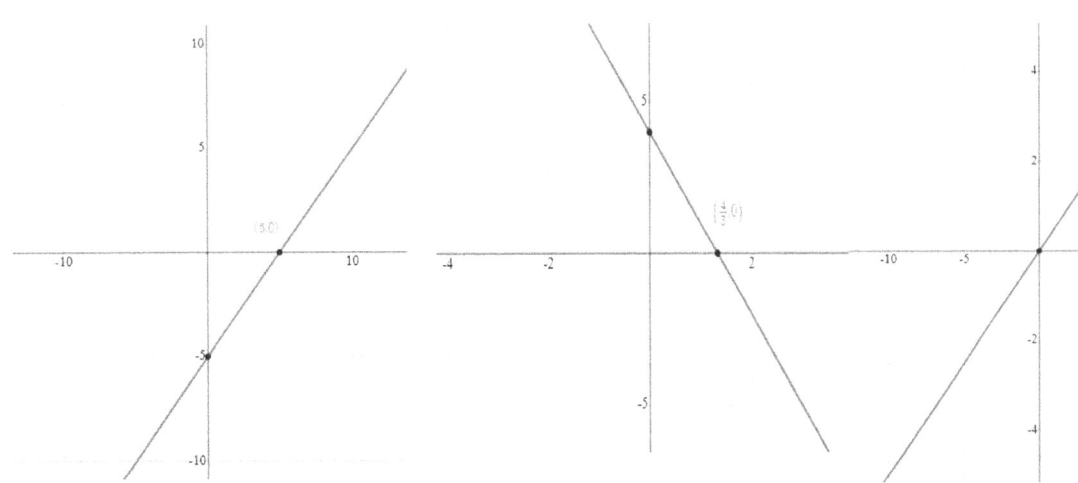

4) $x + y = -4$

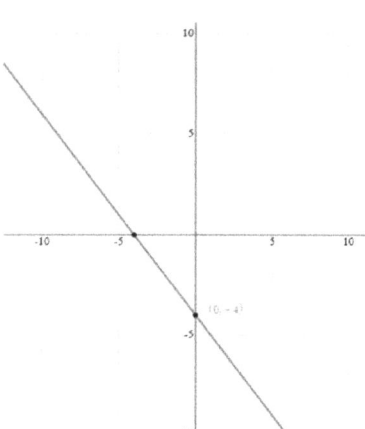

5) $4x + 3y = -2$

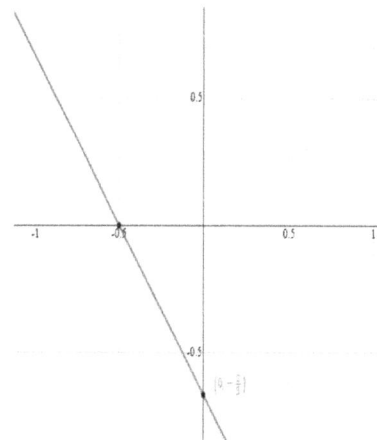

6) $y - 3x + 6 = 0$

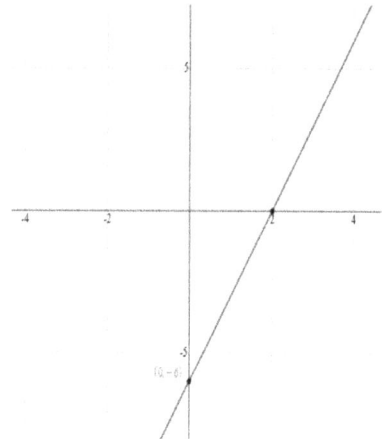

Writing Linear Equations

1) $y = 6x - 46$
2) $y = -3x + 34$
3) $y = x + 10$
4) $y = x - 4$
5) $y = -2x + 23$
6) $y = x - 15$
7) $y = -x + 20$
8) $y = -2x + 21$
9) $y = -3x + 42$
10) $y = -6x + 191$
11) $y = 8x + 57$
12) $y = 4x - 24$
13) $y = 4x - 34$
14) $y = -2x - 4$
15) $y = \frac{7}{6}x - \frac{5}{3}$
16) $y = -\frac{11}{3}x + \frac{28}{3}$
17) $y = 3x + 11$
18) $y = 5x + 15$
19) $y = 7x - 19$
20) $y = -3x - 1$
21) $y = -6x - 11$
22) $y = 5x + 22$

Graphing Linear Inequalities

1) $y > 3x - 5$

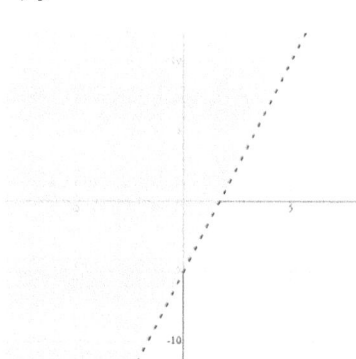

2) $y < 2x + 1$

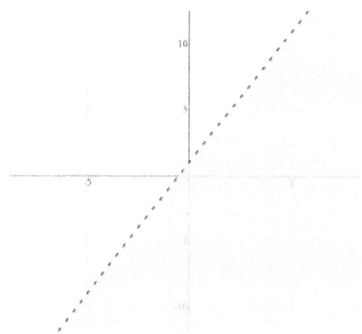

3) $y \leq -4x - 5$

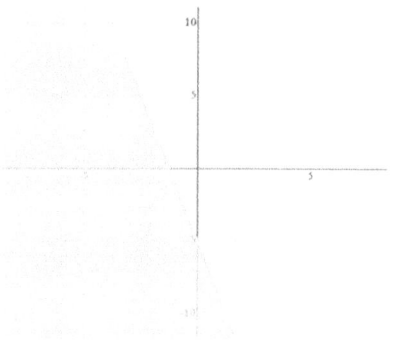

4) $2y \geq 12 + 4x$ 5) $-5y < x - 15$ 6) $3y \geq -9x + 6$

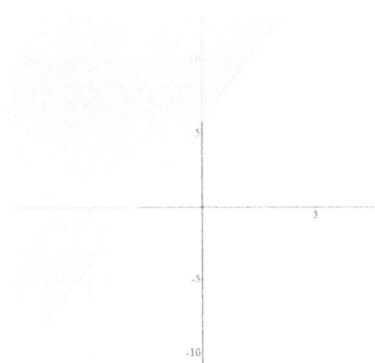 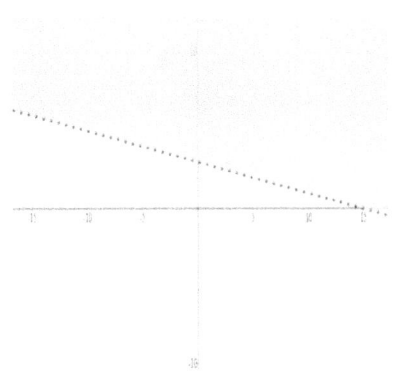 

Write an equation from a graph

1) $y = 4x + 2$ 2) $y = 2x + 6$ 3) $y = 8x$ 4) $y = 3x + 1$

Rate of change

1) 14   2) -10 3) 2

x–intercept and y–intercept

1) $y - intercept = 6$ $x - intercept = 3$

2) $y - intercept = 4$ $x - intercept = -4$

3) $y - intercept = -18$ $x - intercept = 6$

4) $y - intercept = -6$ $x - intercept = -6$

5) $y - intercept = -\frac{8}{5}$ $x - intercept = \frac{8}{7}$

6) $y - intercept = -\frac{12}{5}$ $x - intercept = 3$

7) $y - intercept = -\frac{15}{4}$ $x - intercept = -\frac{5}{4}$

8) $y - intercept =$ undefind $x - intercept = 4$

9) $y - intercept = 4$ $x - intercept =$ undefind

10) $y - intercept = 6$ $x - intercept = 10$

Find the value of b

11) -11 12) 1   13) 3   14) 2

Slope–intercept form

1) $y = 15x + 7$ 4) $y = -\frac{1}{2}x - \frac{3}{2}$ 6) $y = \frac{7}{5}x + 2$ 8) $y = \frac{6}{5}x + \frac{36}{5}$

2) $y = -5x - 12$ 5) $y = \frac{2x}{5} - 3$ 7) $y = 3x + 6$ 9) $y = 12x - 20$

3) $y = -\frac{1}{3}x + 2$

10) $y = -\frac{5}{2}x + 4$ 12) $y = -\frac{12}{7}x - 3$ 14) $y = 2x + 11$ 16) $y = -\frac{3}{2}x - 2$

11) $y = \frac{3}{2}x + 4$ 13) $y = \frac{1}{3}x - \frac{4}{3}$ 15) $y = \frac{3}{4}x - \frac{7}{2}$

Point–slope form

1) $m = 3, (-5, 0)$
2) $m = \frac{1}{4}, (3, -2)$
3) $m = -\frac{5}{2}, (0, -1)$
4) $m = 3, (4, 4)$
5) $m = \frac{6}{10}, (-7, -5)$
6) $m = -2, (0, 6)$
7) $m = -2, (9, 10)$

8) $m = 0, (0, -18)$
9) $m = 6, (-1, -19)$
10) $m = -3, (-2, 14)$
11) $y + 7 = 5(x - 9)$
12) $y - 5 = \frac{1}{2}(x + 2)$
13) $y + 4 = -3x$
14) $y - b = n(x + a)$

15) $y - 2 = 4(x + 8)$
16) $y - 1 = -4(x - 6)$
17) $y - 12 = \frac{1}{6}(x + 7)$
18) $y - 13 = 0$
19) $y - 2 = \frac{1}{7}\left(x + \frac{1}{2}\right)$
20) $y = -5x$

Graphing Line of Equation

1) $y = 3x - 2$

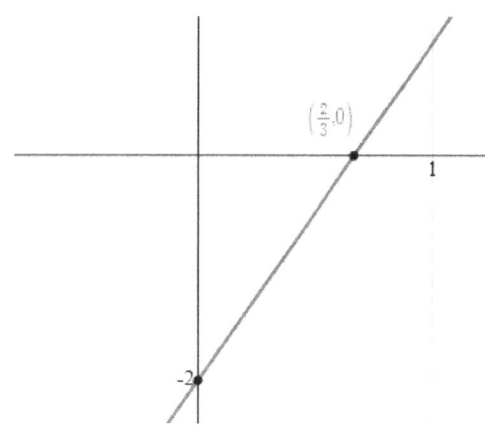

2) $y = -\frac{1}{2}x + \frac{3}{2}$

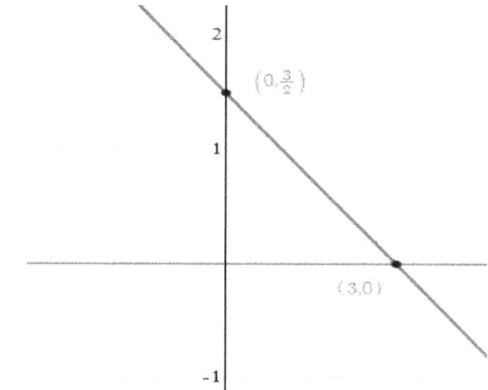

3) $4x - 5y = 12$

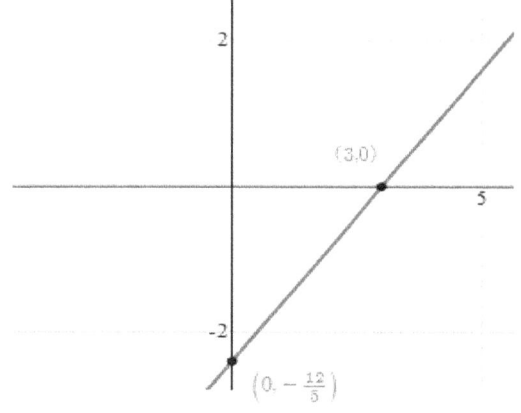

4) $-3x - y = 5$

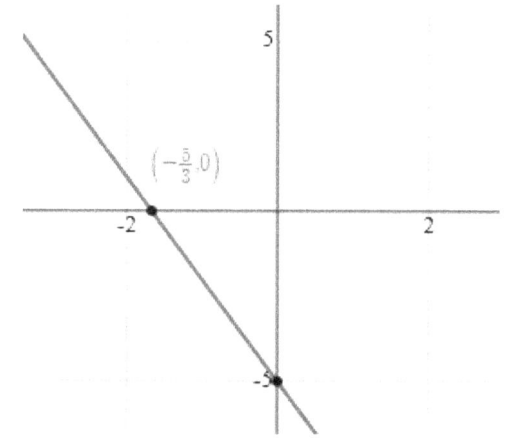

CLEP College Algebra Workbook

Equation of parallel or perpendicular line.

1) $y = -\frac{1}{2}x - 2\frac{1}{2}$

2) $y = x + 5$

3) $y = \frac{1}{4}x + \frac{7}{4}$

4) $y = 2x + 1$

5) $y = 8$

6) $y = -2x + 6$

7) $y = \frac{2}{3}x + 2$

8) $y = 4x + 7$

9) $y = -\frac{1}{5}x - \frac{4}{5}$

10) $y = -\frac{1}{5}x - \frac{18}{5}$

11) $y = \frac{1}{2}x + 2$

12) $y = -\frac{2}{3}x + \frac{5}{12}$

13) $x = 4$

14) $y = -5$

15) $y = -4x - 25$

16) $y = \frac{1}{4}x + \frac{7}{12}$

17) $y = -\frac{4}{7}x - \frac{60}{7}$

18) $y = -x - 3$

19) $y = -\frac{4}{9}x - \frac{106}{9}$

20) $y = -5x + 27$

Equations of horizontal and vertical lines

1) $y = -2$

2) $y = 1$

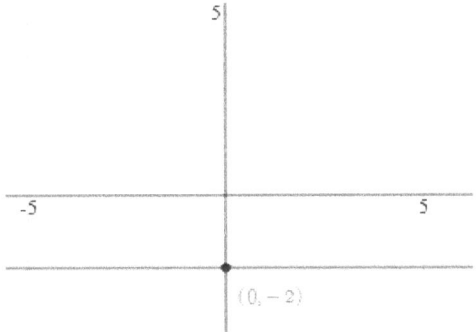
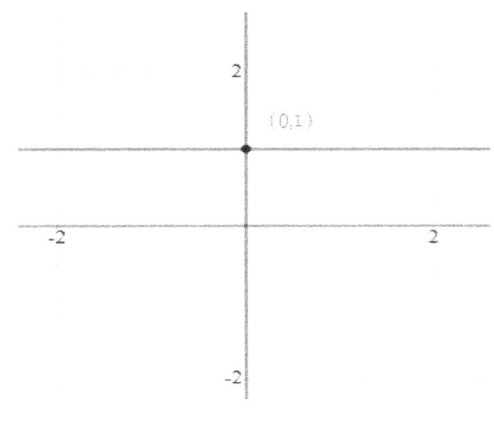

3) $x = -1$

4) $x = 2$

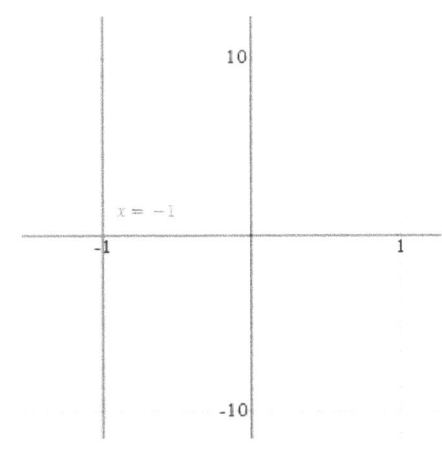
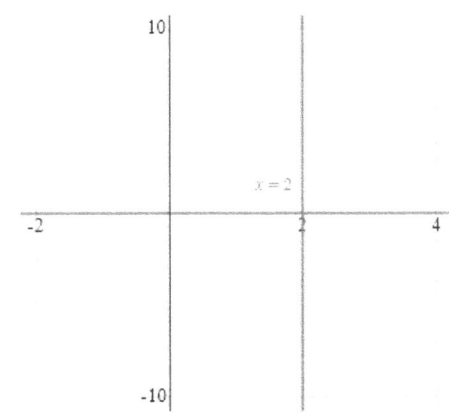

Graphing Absolute Value Equations

1) $y = |x + 4|$

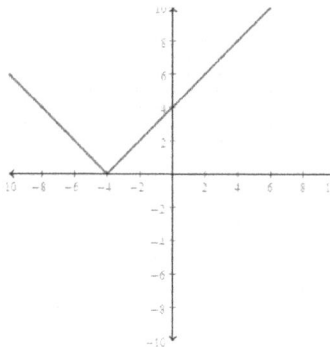

2) $y = |x - 1|$

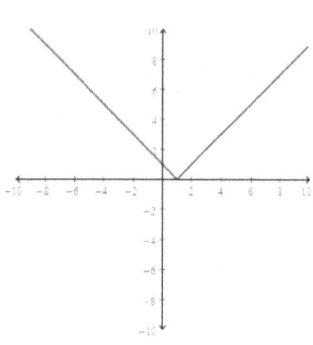

3) $y = -|x| - 1$

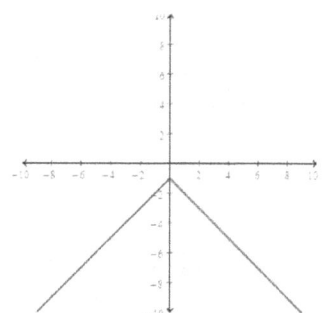

4) $y = |x - 2|$

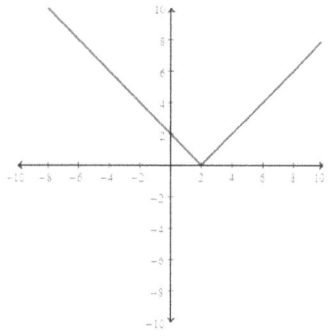

5) $y = -|x - 2|$

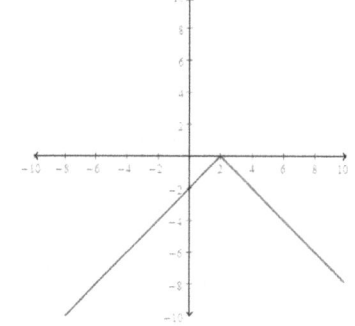

6) $y = -2|2x + 2| + 4$

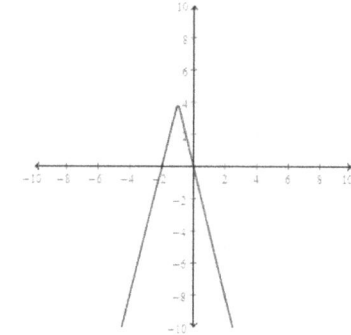

Chapter 5:
Monomials and polynomials

Topics that you will practice in this chapter:

- ✓ GCF of Monomials
- ✓ Factoring Quadratics
- ✓ Factoring by Grouping
- ✓ GCF and Powers of Monomials
- ✓ Writing Polynomials in Standard Form
- ✓ Simplifying Polynomials
- ✓ Adding and Subtracting Polynomials
- ✓ Multiplying a Polynomial and a Monomial
- ✓ Multiplying Binomials
- ✓ Factoring Trinomials
- ✓ Operations with Polynomials

Mathematics is, as it were, a sensuous logic, and relates to philosophy as do the arts, music, and plastic art to poetry. — *K. Shegel*

GCF of Monomials

✏️ **Find the GCF of each set of monomials.**

1) $39x, 30xy$

2) $60a, 56a^2$

3) $18x^2, 54x^2$

4) $36x^2, 21x^3$

5) $20a^2, 30a^2b$

6) $80a^3, 30a^2b$

7) $54x^3, 36x^3$

8) $33x, 44y^2x$

9) $15x^2, 12, 48$

10) $10v^3, 45v^3, 35v$

11) p^2q^2, pqr

12) $15m^2n, 25m^2n^2$

13) $12x^2yz, 3xy^2$

14) $22m^5n^2, 11m^2n^4$

15) $16x^3y, 8x^2$

16) $14ab^5, 7a^2b^2c$

17) $12t^7u^2, 18t^3u^7$

18) $18t, 48t^4$

19) $18r^3t, 26qr^2t^4$

20) $11a^4b^3, 44a^2b^5$

21) $16f, 21ab^2$

22) $12a^2b^2c^2, 20abc$

23) $18ab, 9ab$

24) $22m^5n^2, 11m^2n^4$

25) $4xy, 2x^2$

26) $x^3yz^2, 2x^3yz^3$

27) $140x, 140y^2, 80y^2$

28) $24a, 36a, 24ab^2$

29) $10x^3, 45x^3, 35x$

30) $105a, 30ab, 75a$

Factoring Quadratics

Factor each completely.

1) $x^2 - 16x + 63 =$

2) $m^2 - 9m + 8 =$

3) $p^2 - 5p - 14 =$

4) $2b^2 + 17b + 21 =$

5) $a^2 + 5a + 4 =$

6) $a^2 + 2a - 15 =$

7) $4n^2 + 12n + 9 =$

8) $t^2 + 2t - 19 =$

9) $3x^3 + 21x^2 + 36x =$

10) $x^2 + 5x + 6 =$

11) $9r^2 - 5r - 10 =$

12) $30n^2b - 87nb + 30b =$

13) $7x^2 - 32x - 60 =$

14) $3b^3 - 5b^2 + 2b =$

15) $10m^2 + 89m - 9 =$

16) $4x^3 + 43x^2 + 30x =$

17) $9x^2 + 7 - 56 =$

18) $p^2 - 5p - 14 =$

19) $x^2 - 7x - 18 =$

20) $7x^2 - 31x - 20 =$

21) $6n^2 + 7n - 49 =$

22) $-6x^2 - 25x - 25 =$

23) $6x^2 + 5x - 6 =$

24) $16x^2 + 60x - 100 =$

25) $4x^2 - 35x + 49 =$

26) $5x^2 - 18x + 9 =$

27) $9n^2 + 66n + 21 =$

28) $3x^2 - 8x + 4 =$

29) $6x^2 - 36xy =$

30) $-6x^3 - 23x^2y - 10y^2x =$

31) $9a^2 + 9ab - 4b^2 =$

32) $4x^2 + 4xy - 35y^2 =$

33) $7x^2y - 27xy^2 + 18y^3 =$

34) $-2x^2 + 8xy + 64y^2 =$

35) $25mp^2 - 45mp =$

36) $14b^2 + 142b + 144 =$

37) $5x^2 + 85xy + 350y^2 =$

38) $7x^2 + 9xy =$

Factoring by Grouping

✎ **Factor each completely.**

1) $28xy - 7k - 49x + 4ky =$

2) $7xy - 3n - x + 21ny =$

3) $56n^3 + 64n^2 + 70n + 80 =$

4) $32u^2v - 12u^3m + 48u^4 - 8umv =$

5) $70n^4 + 40n^3 + 28n^2 + 16n =$

6) $45uv - 125bu - 75u^2 + 75bv =$

7) $x^3 + 7x^2 + 6x + 42 =$

8) $6x^3 + 36x^2 + 30x + 180 =$

9) $6m^3 - 30m^2 + 30m - 150 =$

10) $2x^3 - 4x^2 - 10x + 20 =$

11) $24p^3 + 15p^2 - 56p - 35 =$

12) $42mc + 36md - 7n^2c - 6n^2d =$

13) $28x^4 + 112x^2 - 21x^2 - 84x =$

14) $15xw + 18xk + 25yw + 30k =$

15) $56xy - 35x + 16ry - 10r =$

16) $4xy + 6 - x - 24y =$

17) $192x3 + 72x2 + 144x + 54 =$

18) $8x3 - 8x2 + 14x - 14 =$

19) $20x^3 + 5x^2 + 28x + 7 =$

20) $100x^3 + 160x^2 - 60x - 96 =$

GCF and Powers of Monomials

✎ **Find the GCF of each pairs of expressions.**

1) $54x^3, 36x^3$

2) 2) $33x, 44y^2x$

3) $15x^2, 12, 48$

4) 4) $10v^3, 45v^3, 35v$

5) p^2q^2, pqr

6) 6) $15m^2n, 25m^2n^2$

7) $12x^2yz, 3xy^2$

8) 8) $22m^5n^2, 11m^2n^4$

9) $16x^3y, 8x^2$

10) $14ab^5, 7a^2b^2c$

11) $12t^7u^2, 18t^3u^7$

12) 12) $18t, 48t^4$

13) $18r^3t, 26qr^2t^4$

14) 14) $11a^4b^3, 44a^2b^5$

15) $16f, 21ab^2$

16) 16) $12a^2b^2c^2, 20abc$

17) $18ab, 9ab$

18) 18) $22m^5n^2, 11m^2n^4$

19) $4xy, 2x^2$

20) 20) $x^3yz^2, 2x^3yz^3$

✎ **Simplify.**

21) $(3x^4)^7$

22) $(4y^2 2y^3 y)^2$

23) $(3x^2 2x^2)^3$

24) $(8x^4y^3)^6$

25) $(3y^2 5y^2)^2$

26) $(6x^3y)^3$

27) $(8x^2 x^2 3n)^2$

28) $(7xy^6)^3$

29) $(9x^3y^2)^4$

30) $(10y^3y^2)^3$

31) $(6x^2x^6)^3$

32) $(3x^7 4x^3 k^2)^2$

33) $(4y^5 4y^2)^2$

34) $(5x 2x^3)^3$

35) $(4y^3)^3$

36) $(y^3y^3y^2)^3$

37) $(4y^2y)^3$

38) $(6xy^6)^3$

Writing Polynomials in Standard Form

✎ **Write each polynomial in standard form.**

1) $9x - 7x =$

2) $-6 + 15x - 15x =$

3) $3x^2 - 11x^3 =$

4) $18 + 19x^3 - 14 =$

5) $3x^2 + 9x - 4x^5 =$

6) $-7x^3 + 12x^7 =$

7) $9x + 6x^2 - 2x^6 =$

8) $-5x^3 + x - 9x^4 =$

9) $8x^2 + 34 - 21x =$

10) $8 - 7x + 11x^4 =$

11) $25x^3 + 45x - 13x^4 =$

12) $17 + 9x^2 - 2x^3 =$

13) $18x^2 - 8x + 8x^3 =$

14) $9x^4 - 4x^2 - 10x^5 =$

15) $-41 + 7x^2 - 8x^4 =$

16) $8x^2 - 7x^5 + 3x^3 - 12 =$

17) $4x^2 - 9x^5 + 12 - 8x^4 =$

18) $-2x^5 + 6x - 9x^2 - 7x =$

19) $14x^5 + 7x^4 - 8x^5 - 8x^2 =$

20) $2x^3 - 15x^4 + 9x^3 + 3x^8 =$

21) $7x^4 - 16x^5 - 9x^2 + 10x^4 =$

22) $5x^2 + 6x^5 + 37x^3 - 9x^5 =$

23) $3x(2x + 5 - 6x^2) =$

24) $12x(x^6 + 2x^3) =$

25) $6x(x^2 + 8x + 4) =$

26) $8x(3 - 2x + 4x^3) =$

27) $7x(2x^3 - 2x^2 + 2) =$

28) $5x(5x^5 + 4x^4 - 1) =$

29) $x(4x^3 + 52x^4 + 2x) =$

30) $6x(3x - 4x^4 + 7x^2) =$

Simplifying Polynomials

✏️ **Simplify each expression.**

1) $3(x - 12) =$

2) $5x(2x - 4) =$

3) $7x(5x - 1) =$

4) $6x(3x + 2) =$

5) $5x(2x - 7) =$

6) $9x(x + 8) =$

7) $(3x - 8)(x - 3) =$

8) $(x - 9)(3x + 4) =$

9) $(x - 8)(x - 5) =$

10) $(3x + 4)(3x - 4) =$

11) $(5x - 8)(5x - 2) =$

12) $7x^2 + 7x^2 - 6x^4 =$

13) $5x - 2x^2 + 7x^3 + 10 =$

14) $8x + 2x^2 - 5x^3 =$

15) $15x + 4x^5 - 8x^2 =$

16) $-4x^2 + 7x^5 + 11x^4 =$

17) $-14x^2 + 8x^3 - 2x^4 + 5x =$

18) $14 - 5x^2 + 6x^2 - 10x^3 + 17 =$

19) $x^2 - 9x + 2x^3 + 15x - 10x =$

20) $14 - 8x^2 + 4x^2 - 9x^3 + 1 =$

21) $-4x^5 + 2x^4 - 18x^2 + 2x^5 =$

22) $(3x^3 - 5) + (3x^3 - 2x^3) =$

23) $4(3x^5 - 3x^3 - 6x^5) =$

24) $-4(x^5 + 8) - 4(12 - x^5) =$

25) $7x^2 - 9x^3 - 2x + 14 - 5x^2 =$

26) $10 - 5x^2 + 3x^2 - 4x^3 + 4 =$

27) $(8x^2 - 2x) - (5x - 5 - 4x^2) =$

28) $4x^4 - 8x^3 - x(3x^2 + 5x) =$

29) $4x + 8x^2 - 10 - 2(x^2 - 1) =$

30) $5 - 3x^2 + (6x^4 - 2x^2 + 8x^4) =$

31) $-(x^5 + 8) - 7(4 + x^5) =$

32) $(4x^3 - x) - (x - 6x^3) =$

Adding and Subtracting Polynomials

✎ **Add or subtract expressions.**

1) $(-x^3 - 3) + (4x^3 + 2) =$

2) $(3x^2 + 4) - (6 - x^2) =$

3) $(x^3 + 4x^2) - (5x^3 + 15) =$

4) $(3x^3 - 2x^2) + (2x^2 - x) =$

5) $(10x^3 + 14x) - (14x^3 + 7) =$

6) $(5x^2 - 7) + (3x^2 + 7) =$

7) $(9x^3 + 4) - (10 - 5x^3) =$

8) $(x^2 + 2x^3) - (2x^3 + 5) =$

9) $(8x^2 - x) + (5x - 4x^2) =$

10) $(17x + 10) - (2x + 10) =$

11) $(12x^4 - 4x) - (x - 3x^4) =$

12) $(3x - x^4) - (7x^4 + 8x) =$

13) $(7x^3 - 6x^5) - (4x^5 - 2x) =$

14) $(x^3 - 7) + (4x^3 + 8x^5) =$

15) $(6x^2 + 5x^4) - (x^4 - 9x^2) =$

16) $(-4x^2 - 4x) + (7x - 8x^2) =$

17) $(x - 6x^4) - (15x^4 + 2x) =$

18) $(4x - 3x^4) - (2x^4 - 3x^3) =$

19) $(7x^3 - 7) + (6x^3 - 6x^2) =$

20) $(9x^5 + 7x^4) - (x^4 - 5x^5) =$

21) $(-4x^2 + 11x^4 + 2x^3) + (20x^3 + 4x^4 + 12x^2) =$

22) $(5x^2 - 5x^4 - 5x) - (-4x^2 - 5x^4 + 5x) =$

23) $(12x + 36x^3 - 10x^4) + (20x^3 + 10x^4 - 7x) =$

24) $(2x^5 - 4x^3 - 5x) - (2x^2 + 7x^3 - 2x) =$

25) $(14x^3 - 4x^5 - x) - (-4x^3 - 12x^5 + 9x) =$

26) $(-5x^2 + 12x^4 + x^3) + (10x^3 + 17x^4 + 7x^2) =$

WWW.MathNotion.Com

Multiplying a Polynomial and a Monomial

✍ **Find each product.**

1) $2x(x+4) =$

2) $3(8-x) =$

3) $5x(3x+4) =$

4) $x(-2x+5) =$

5) $7x(3x-3) =$

6) $3(2x-5y) =$

7) $6x(7x-3) =$

8) $x(12x+5y) =$

9) $5x(x+6y) =$

10) $11x(4x+5y) =$

11) $8x(4x+2) =$

12) $12x(x-15y) =$

13) $9x(5x-3y) =$

14) $8x(5x-2y+5) =$

15) $9x(2x^2+7y^2) =$

16) $8x(9x+6y) =$

17) $2(3x^5-2y^5) =$

18) $4x(-x^2y+2y) =$

19) $-3(2x^3-3xy+9) =$

20) $2(x^2-2xy-4) =$

21) $7x(4x^3-xy+2x) =$

22) $-9x(-2x^3-2x+7xy) =$

23) $6(x^2+3xy-8y^2) =$

24) $5x(7x^3-x+8) =$

25) $7(x^{24}-4x-6) =$

26) $x^2(-3x^3+4x+7) =$

27) $x^2(2x^3+10-5x) =$

28) $4x^4(3x^3-2x+8) =$

29) $5x^2(x^4-5xy+2y^3) =$

30) $4x^2(7x^4-2x+11) =$

31) $7x^3(3x^3+5x-7) =$

32) $4x(x^2-8xy+7y^3) =$

Multiplying Binomials

✏️ **Find each product.**

1) $(x + 5)(x + 1) =$

2) $(x - 3)(x + 7) =$

3) $(x - 1)(x - 9) =$

4) $(x + 3)(x + 8) =$

5) $(x - 4)(x - 11) =$

6) $(x + 5)(x + 6) =$

7) $(x - 8)(x + 7) =$

8) $(x - 3)(x - 2) =$

9) $(x + 8)(x + 11) =$

10) $(x - 3)(x + 5) =$

11) $(x + 8)(x + 8) =$

12) $(x + 2)(x + 7) =$

13) $(x - 9)(x + 4) =$

14) $(x - 10)(x + 10) =$

15) $(x + 24)(x + 2) =$

16) $(x + 9)(x + 13) =$

17) $(x - 7)(x + 7) =$

18) $(x - 5)(x + 2) =$

19) $(3x + 4)(x + 5) =$

20) $(x - 8)(5x + 2) =$

21) $(x - 9)(4x + 9) =$

22) $(2x - 7)(3x - 2) =$

23) $(x - 4)(x + 11) =$

24) $(5x - 6)(2x + 4) =$

25) $(4x - 9)(x + 7) =$

26) $(8x - 5)(2x + 2) =$

27) $(3x + 9)(7x + 4) =$

28) $(6x - 8)(4x + 4) =$

29) $(4x + 5)(5x - 8) =$

30) $(8x - 1)(8x + 4) =$

31) $(9x + 4)(3x - 6) =$

32) $(4x^2 + 12)(4x^2 - 12) =$

Factoring Trinomials

✏️ **Factor each trinomial.**

1) $x^2 + 12x + 35 =$

2) $x^2 - 8x + 12 =$

3) $x^2 + 11x + 10 =$

4) $x^2 - 12x + 27 =$

5) $x^2 - 16x + 15 =$

6) $x^2 - 13x + 40 =$

7) $x^2 + 15x + 44 =$

8) $x^2 + x - 72 =$

9) $x^2 - 81 =$

10) $x^2 - 17x + 70 =$

11) $x^2 + 8x - 48 =$

12) $x^2 + 5x - 104 =$

13) $x^2 - 7x - 18 =$

14) $x^2 + 22x + 121 =$

15) $3x^2 - 3x - 36 =$

16) $2x^2 - 35x + 75 =$

17) $14x^2 + 11x - 15 =$

18) $8x^2 - 12x - 20 =$

19) $15x^2 + 16x + 4 =$

20) $24x^2 + 2x - 1 =$

✏️ **Calculate each problem.**

21) The area of a rectangle is $x^2 - 3x - 40$. If the width of rectangle is $x - 8$, what is its length? _____

22) The area of a parallelogram is $12x^2 + 7x - 10$ and its height is $4x + 5$. What is the base of the parallelogram? _____

23) The area of a rectangle is $10x^2 - 43x + 28$. If the width of the rectangle is $5x - 4$, what is its length? _____

Operations with Polynomials

✎ **Find each product.**

1) $2(4x + 1) = $ _____

2) $5(2x + 7) = $ _____

3) $4(6x - 5) = $ _____

4) $-4(7x - 8) = $ _____

5) $3x^2(8x + 4) = $ _____

6) $6x^2(2x - 9) = $ _____

7) $5x^3(-x + 4) = $ _____

8) $-5x^4(4x - 9) = $ _____

9) $6(x^2 + 7x - 3) = $ _____

10) $4(3x^2 - 2x + 6) = $ _____

11) $9(3x^2 + 8x + 2) = $ _____

12) $7x(x^2 + 5x + 3) = $ _____

13) $(7x + 2)(2x - 5) = $ _____

14) $(8x + 5)(3x - 8) = $ _____

15) $(4x + 2)(6x - 1) = $ _____

16) $(5x - 4)(5x + 9) = $ _____

✎ **Calculate each problem.**

17) The measures of two sides of a triangle are $(2x + 8y)$ and $(5x - 3y)$. If the perimeter of the triangle is $(11x + 6y)$, what is the measure of the third side? _____

18) The height of a triangle is $(8x + 2)$ and its base is $(2x - 6)$. What is the area of the triangle? _____

19) One side of a square is $(4x + 3)$. What is the area of the square? _____

20) The length of a rectangle is $(7x - 9y)$ and its width is $(13x + 9y)$. What is the perimeter of the rectangle? _____

21) The side of a cube measures $(x + 2)$. What is the volume of the cube? _____

22) If the perimeter of a rectangle is $(24x + 10y)$ and its width is $(4x + 3y)$, what is the length of the rectangle? _____

Answers of Worksheets – Chapter 5

GCF of Monomials

1) $3x$
2) 4
3) $18x^2$
4) $3x^2$
5) $10a^2$
6) $10a^2$
7) $18x^3$
8) $11x$
9) 3
10) $5v$
11) pq
12) $5m^2n$
13) $3xy$
14) $11m^2n$
15) $8x^2$
16) $7ab^2$
17) $6t^3u^2$
18) $8t$
19) $2r^2t$
20) $11a^2b^3$
21) no
22) $4abc$
23) $9ab$
24) $11m^2n^2$
25) $2x$
26) x^3yz^2
27) 20
28) $12a$
29) $5x$
30) $15a$

Factoring Quadratics

1) $(x-9)(x-7)$
2) $(m-1)(m-8)$
3) $(p+2)(p-7)$
4) $(2b+3)(b+7)$
5) a^2+5a+4
6) $a^2+2a-15$
7) $4n^2+12n+9$
8) $t^2+2t-19$
9) $3x^3+21x^2+36x$
10) x^2+5x+6
11) $9r^2-5r-10$
12) $30n^2b\ 87nb+30b$
13) $7x^2-32x-60$
14) $3b^3-5b^2+2b$
15) $10m^2+89m-9$
16) $4x^3+43x^2+30x$
17) $9x^2+7x-56$
18) $p^2-5p-14$
19) $x^2-7x-18$
20) $7x^2-31x-20$
21) $(3n-7)(2x+7)$
22) $-(2x+5)(3x+5)$
23) $(2x+3)(3x-2)$
24) $4(x+5)(4x-5)$
25) $(x-7)(4x-7)$
26) $(5x-3)(x-3)$
27) $3(3n+1)(n+7)$
28) $(3x-2)(x-2)$
29) $6x(x-6y)$
30) $-x(2x+y)(3x+10y)$
31) $(3a+4b)(3a-b)$
32) $(2x+7y)(2x-5y)$
33) $y(7x-6y)(x-3y)$
34) $-2(x-8y)(x+4y)$
35) $5mp(5p-9)$
36) $2(7b+8)(b+9)$

37) $5(x + 10y)(x + 7y)$

38) $x(7x + 9y)$

Factoring by Grouping

1) $(7x + k)(4y - 7)$
2) $(x + 3n)(7y - 1)$
3) $2(4n^2 + 5)(7n + 8)$
4) $4u(4u - m)(2v + 3u^2)$
5) $2n(5n^2 + 2)(7n + 4)$
6) $5(3u + 5b)(3v - 5u)$
7) $(x^2 + 6)(x + 7)$
8) $6(x^2 + 5)(x + 6)$
9) $6(m^2 + 5)(m - 5)$
10) $2(x^2 - 5)(x - 2)$

11) $(3p^2 - 7)(8p + 5)$
12) $(6m - n^2)(7c + 6d)$
13) $7x(4x^2 - 3)(x + 4)$
14) $(3x + 5y)(5w + 6k)$
15) $(7x + 2r)(8y - 5)$
16) $(x - 6)(4y - 1)$
17) $(4x^2 + 1)(3x - 5)$
18) $2(4x^2 + 7)(x - 1)$
19) $(5x^2 + 7)(4x + 1)$
20) $4(5x^2 - 3)(5x + 8)$

GCF and Powers of monomials

1) $10x^2$
2) 4
3) 10
4) $7xy$
5) $6y7x$
6) $3x$
7) 3
8) $7xy^2$
9) $18x^2$
10) $(15)x$
11) 27
12) 3

13) $20x$
14) $2x^2y$
15) xy
16) $2x^2y^2$
17) $6xy^4$
18) $5x^3y$
19) $2187x^{28}$
20) $72x^{12}$
21) $216x^{12}$
22) $262144x^{24}y^{18}$
23) $225y^{10}$
24) $216x^9y^3$

25) $576x^8n^2$
26) $343x^3y^{18}$
27) $6561x^{12}y^8$
28) $1000y^{15}$
29) $396x^{24}$
30) $144x^{100}k^4$
31) $256y^{14}$
32) $1000x^{12}$
33) $64y^9$
34) $27y^{18}$
35) $64y^9$
36) $216x^3y^{18}$

Writing Polynomials in Standard Form

1) $2x$
2) -6
3) $-11x^3 + 3x^2$
4) $19x^4 + 4$

5) $-4x^5 + 3x^2 + 9x$
6) $12x^7 - 7x^3$
7) $-2x^6 + 6x^2 + 9x$
8) $-9x^4 - 5x^3 + x$

9) $8x^2 - 21x + 34$
10) $11x^4 - 7x + 8$
11) $-13x^4 + 25x^3 + 45x$
12) $-2x^3 + 9x^2 + 17$
13) $8x^3 + 18x^2 - 8x$
14) $-10x^5 - 9x^4 - 4x^2$
15) $-8x^4 + 7x^2 - 41$
16) $-7x^5 + 3x^3 + 8x^2 - 12$
17) $-9x^5 - 8x^4 + 4x^2 + 12$
18) $-2x^5 - 9x^2 - x$
19) $6x^5 + 7x^4 - 8x^2$
20) $3x^8 - 15x^4 + 11x^2$
21) $-16x^5 + 17x^4 - 9x^2$
22) $-3x^5 + 37x^3 + 5x^2$
23) $-18x^3 + 6x^2 + 15x$
24) $12x^7 + 24x^4$
25) $6x^3 + 48x^2 + 24x$
26) $32x^4 - 16x^2 + 24x$
27) $14x^4 - 14x^3 + 14x$
28) $25x^6 + 20x^5 - 5x$
29) $52x^5 + 4x^4 + 2x^2$
30) $-24x^5 + 42x^3 + 18x^2$

Simplifying Polynomials

1) $3x - 36$
2) $10x^2 - 20x$
3) $35x^2 - 7x$
4) $18x^2 + 12x$
5) $10x^2 - 35x$
6) $9x^2 + 72x$
7) $3x^2 - 17x + 24$
8) $3x^2 - 23x - 36$
9) $x^2 - 13x + 40$
10) $9x^2 - 16$
11) $25x^2 - 50x + 16$
12) $-6x^4 + 14x^2$
13) $7x^3 - 2x^2 + 5x + 10$
14) $-5x^3 + 2x^2 + 8x$
15) $4x^5 - 8x^2 + 15x$
16) $7x^5 + 11x^4 - 4x^2$
17) $-2x^4 + 8x^3 - 14x^2 + 5x$
18) $-10x^3 + x^2 + 31$
19) $2x^3 + x^2 - 4x$
20) $-9x^3 - 4x^2 + 15$
21) $-2x^5 + 2x^4 - 18x^2$
22) $4x^3 - 5$
23) $-12x^5 - 12x^3$
24) -80
25) $-9x^3 + 2x^2 - 2x + 14$
26) $-4x^3 - 2x^2 + 14$
27) $12x^2 - 7x + 5$
28) $4x^4 - 11x^3 - 5x^2$
29) $6x^2 + 4x - 8$
30) $14x^4 - 5x^2 + 5$
31) $-8x^5 - 36$
32) $10x^3 - 2x$

Adding and Subtracting Polynomials

1) $3x^2 - 1$
2) $4x^2 - 2$
3) $-4x^3 + 4x^2 - 15$
4) $3x^3 - x$
5) $-4x^3 + 14x - 7$
6) $8x^2$

CLEP College Algebra Workbook

7) $14x^3 - 6$
8) $x^2 - 5$
9) $4x^2 + 4x$
10) $15x$
11) $15x^4 - 5x$
12) $-8x^4 - 5x$
13) $-10x^5 + 7x^3 + 2x$

14) $5x^5 + 5x^3 - 7$
15) $4x^4 + 15x^2$
16) $-12x^2 + 3x$
17) $-21x^4 - x$
18) $-5x^4 + 3x^3 + 4x$
19) $13x^3 - 6x^2 - 7$
20) $14x^5 + 6x^4$

21) $15x^4 + 22x^3 + 8x^2$
22) $9x^2 - 10x$
23) $56x^3 + 5x$
24) $2x^5 - 11x^3 - 2x^2 - 3x$
25) $8x^5 + 18x^3 - 10x$
26) $29x^4 + 11x^3 + 2x^2$

Multiplying a Polynomial and a Monomial

1) $2x^2 + 8x$
2) $-3x + 24$
3) $15x^2 + 20x$
4) $-2x^2 + 5x$
5) $21x^2 - 21x$
6) $6x - 15y$
7) $42x^2 - 18x$
8) $12x^2 + 5xy$
9) $5x^2 + 30xy$
10) $44x^2 + 55xy$
11) $32x^2 + 16x$
12) $12 - 180xy$
13) $45x^2 - 27xy$
14) $40x^2 - 16xy + 40x$
15) $18x^3 + 63xy^2$
16) $72x^2 + 48xy$

17) $6x^5 - 2y^5$
18) $-4x^3y + 8xy$
19) $-6x^3 + 9xy - 27$
20) $2x^2 - 4xy - 8$
21) $28x^4 - 7x^2y + 14x^2$
22) $18x^4 + 18x^2 - 63x^2y$
23) $6x^2 + 18xy - 48y^2$
24) $35x^4 - 5x^2 + 40x$
25) $7x^{24} - 28x - 42$
26) $-3x^5 + 4x^3 + 7x^2$
27) $2x^5 - 5x^3 + 10x^2$
28) $12x^7 - 8x^5 + 32x^4$
29) $5x^6 - 25x^3y + 10x^2y^3$
30) $28x^6 - 8x^3 + 44x^2$
31) $21x^6 + 35x^4 - 49x^3$
32) $4x^3 - 32x^2y + 28xy^3$

Multiplying Binomials

1) $x^2 + 6x + 5$
2) $x^2 + 4x - 21$
3) $x^2 - 10x + 9$
4) $x^2 + 11x + 24$
5) $x^2 - 15x + 44$
6) $x^2 + 11x + 30$

7) $x^2 - x - 56$
8) $x^2 - 5x + 6$
9) $x^2 + 19x + 88$
10) $x^2 + 2x + 15$
11) $x^2 + 16x + 64$
12) $x^2 + 9x + 14$

WWW.MathNotion.Com

13) $x^2 - 5x - 36$
14) $x^2 - 100$
15) $x^2 + 26x + 48$
16) $x^2 + 22x + 117$
17) $x^2 - 49$
18) $x^2 - 3x - 10$
19) $3x^2 + 19x + 20$
20) $5x^2 - 38x - 16$
21) $4x^2 - 27x - 81$
22) $6x^2 - 25x + 14$

23) $x^2 + 7x - 44$
24) $10x^2 + 8x - 24$
25) $4x^2 + 19x - 63$
26) $16x^2 + 6x - 10$
27) $21x^2 + 75x + 36$
28) $24x^2 - 8x - 32$
29) $20x^2 - 7x - 40$
30) $64x^2 + 24x - 4$
31) $27x^2 - 42x - 24$
32) $16x^4 - 144$

Factoring Trinomials

1) $(x + 5)(x + 7)$
2) $(x - 2)(x - 6)$
3) $(x + 1)(x + 10)$
4) $(x - 9)(x - 3)$
5) $(x - 1)(x - 15)$
6) $(x - 5)(x - 8)$
7) $(x + 4)(x + 11)$
8) $(x + 9)(x - 8)$

9) $(x - 9)(x + 9)$
10) $(x - 7)(x - 10)$
11) $(x - 4)(x + 12)$
12) $(x - 8)(x + 13)$
13) $(x + 2)(x - 9)$
14) $(x + 11)(x + 11)$
15) $(3x + 9)(x - 4)$
16) $(x - 15)(2x - 5)$

17) $(7x - 5)(2x + 3)$
18) $(2x - 5)(4x + 4)$
19) $(3x + 2)(5x + 2)$
20) $(6x - 1)(4x + 1)$
21) $(x + 5)$
22) $(3x - 2)$
23) $(2x - 7)$

Operations with Polynomials

1) $8x + 2$
2) $10x + 35$
3) $24x - 20$
4) $-28x + 32$
5) $24x^3 + 12x^2$
6) $12x^3 - 54x^2$
7) $-5x^4 + 20x^3$
8) $-20x^5 + 45x^4$

9) $6x^2 + 42x - 18$
10) $12x^2 - 8x + 24$
11) $27x^2 + 72x + 18$
12) $7x^3 + 35x^2 + 21x$
13) $14x^2 - 31x - 10$
14) $24x^2 - 49x - 40$
15) $24x^2 + 8x - 2$
16) $25x^2 + 25x - 36$

17) $(4x + y)$
18) $8x^2 - 22x - 6$
19) $16x^2 + 24x + 9$
20) $40x$
21) $x^3 + 6x^2 + 12x + 8$
22) $(8x + 2y)$

Chapter 6: Functions Operations and Quadratic

Topics that you will practice in this chapter:

- ✓ Evaluating Functions
- ✓ Adding and Subtracting Functions
- ✓ Multiply and Dividing Functions
- ✓ Composition of Functions
- ✓ Solving Quadratic Equations
- ✓ Quadratic Formula and Discriminant
- ✓ Quadratic Inequalities
- ✓ Graphing Quadratic Functions
- ✓ Domain and Range of Radical Functions
- ✓ Solving Radical Equations

It's fine to work on any problem, so long as it generates interesting mathematics along the way – even if you don't solve it at the end of the day." – Andrew Wiles

Evaluating Function

✎ **Write each of following in function notation.**

1) $h = -8x + 9$

2) $k = 5a - 21$

3) $d = 14t$

4) $y = \frac{3}{17}x - \frac{9}{17}$

5) $m = 18n - 94$

6) $c = p^2 - 7p + 15$

✎ **Evaluate each function.**

7) $f(x) = 6x - 7$, find $f(-3)$

8) $g(x) = \frac{1}{10}x + 6$, find $f(5)$

9) $h(x) = -2x + 15$, find $f(8)$

10) $f(x) = -3x + 8$, find $f(-2)$

11) $f(a) = 12a - 9$, find $f(0)$

12) $h(x) = 18 - 5x$, find $f(-4)$

13) $g(n) = 7n - 5$, find $f(5)$

14) $f(x) = -9x - 2$, find $f(3)$

15) $k(n) = -12 + 4.5n$, find $f(2)$

16) $f(x) = -1.5x + 2.5$, find $f(-6)$

17) $g(n) = \frac{16n-8}{6n}$, find $g(2)$

18) $g(n) = \sqrt{5n} - 2$, find $g(5)$

19) $h(x) = x^{-1} - 6$, find $h(\frac{1}{9})$

20) $h(n) = n^{-3} + 4$, find $h(\frac{1}{2})$

21) $h(n) = n^2 - 5$, find $h(\frac{4}{5})$

22) $h(n) = n^3 - 8$, find $h(-\frac{1}{3})$

23) $h(n) = 4n^2 - 42$, find $h(-4)$

24) $h(n) = -5n^2 - 9n$, find $h(7)$

25) $g(n) = \sqrt{4n^2} - \sqrt{5n}$, find $g(5)$

26) $h(a) = \frac{-15a+7}{3a}$, find $h(-b)$

27) $k(a) = 8a - 9$, find $k(a - 3)$

28) $h(x) = \frac{1}{6}x + 7$, find $h(-12x)$

29) $h(x) = 8x^2 + 10$, find $h(\frac{x}{2})$

30) $h(x) = x^4 - 8$, find $h(-2x)$

Adding and Subtracting Functions

✎ **Perform the indicated operation.**

1) $f(x) = 2x + 3$

   $g(x) = x + 4$

   Find $(f - g)(2)$

2) $g(a) = -3a - 8$

   $f(a) = -4a - 12$

   Find $(g - f)(-2)$

3) $h(t) = 7t + 5$

   $g(t) = 3t + 11$

   Find $(h - g)(t)$

4) $g(a) = -5a - 3$

   $f(a) = 3a^2 + 4$

   Find $(g - f)(x)$

5) $g(x) = \frac{2}{7}x - 10$

   $h(x) = \frac{5}{7}x + 10$

   Find $g(14) - h(14)$

6) $h(3) = \sqrt{7x} - 2$

   $g(x) = \sqrt{7x} + 2$

   Find $(h + g)(7)$

7) $f(x) = x^{-3}$

   $g(x) = x^2 + \frac{4}{x}$

   Find $(f - g)(-1)$

8) $h(n) = n^2 + 8$

   $g(n) = -n + 5$

   Find $(h - g)(a)$

9) $g(x) = -2x^2 - 3 - x$

   $f(x) = 7 + x$

   Find $(g - f)(2x)$

10) $g(t) = 4t - 9$

    $f(t) = -t^2 + 5$

    Find $(g + f)(-z)$

11) $f(x) = 3x + 9$

    $g(x) = -4x^2 + 2x$

    Find $(f - g)(-x^2)$

12) $f(x) = -9x^3 - 4x$

    $g(x) = 4x + 12$

    Find $(f + g)(3x^2)$

Multiplying and Dividing Functions

✎ **Perform the indicated operation.**

1) $g(x) = -2x - 5$

   $f(x) = 3x + 4$

   Find $(g.f)(2)$

2) $f(x) = 3x$

   $h(x) = -2x + 5$

   Find $(f.h)(-3)$

3) $g(a) = 5a - 3$

   $h(a) = a - 7$

   Find $(g.h)(-3)$

4) $f(x) = x - 4$

   $h(x) = 4x - 3$

   Find $(\frac{f}{h})(4)$

5) $f(x) = 9a^2$

   $g(x) = 5 + 4a$

   Find $(\frac{f}{g})(3)$

6) $g(a) = \sqrt{5a} + 7$

   $f(a) = (-a)^2 + 3$

   Find $(\frac{g}{f})(5)$

7) $g(t) = t^2 + 5$

   $h(t) = 2t - 5$

   Find $(g.h)(-3)$

8) $g(n) = n^2 + 2n - 4$

   $h(n) = -n + 6$

   Find $(g.h)(1)$

9) $g(a) = (a - 7)^3$

   $f(a) = a^2 + 8$

   Find $(\frac{g}{f})(7)$

10) $g(x) = -x^2 + \frac{4}{5}x + 10$

    $f(x) = x^2 - 3$

    Find $(\frac{g}{f})(5)$

11) $f(x) = x^3 - 3x^2 + 9$

    $g(x) = x - 4$

    Find $(f.g)(x)$

12) $f(x) = 3x - 5$

    $g(x) = x^2 - 4x$

    Find $(f.g)(x^2)$

Composition of Functions

Using $f(x) = 2x - 5$ **and** $g(x) = -2x$, **find:**

1) $f(g(0)) =$

2) $f(g(-1)) =$

3) $g(f(1)) =$

4) $g(f(3)) =$

5) $f(g(-2)) =$

6) $g(f(5)) =$

Using $f(x) = -\frac{1}{4}x + \frac{3}{4}$ **and** $g(x) = x^2$, **find:**

7) $g(f(4)) =$

8) $g(f(3)) =$

9) $g(g(2)) =$

10) $f(f(1)) =$

11) $g(f(-1)) =$

12) $g(f(7)) =$

Using $f(x) = -5x + 2$ **and** $g(x) = x + 3$, **find:**

13) $g(f(0)) =$

14) $f(f(2)) =$

15) $f(g(3)) =$

16) $f(g(-3)) =$

17) $g(f(-5)) =$

18) $f(f(x)) =$

Using $f(x) = \sqrt{x + 9}$ **and** $g(x) = x - 9$, **find:**

19) $f(g(9)) =$

20) $g(f(-8)) =$

21) $f(g(18)) =$

22) $f(f(-5)) =$

23) $g(f(7)) =$

24) $g(g(8)) =$

Quadratic Equation

✎ **Multiply.**

1) $(x-2)(x+8) =$ _____

2) $(x+1)(x+9) =$ _____

3) $(x-5)(x+6) =$ _____

4) $(x+7)(x-3) =$ _____

5) $(x-9)(x-8) =$ _____

6) $(4x+2)(x-4) =$ _____

7) $(3x-6)(x+4) =$ _____

8) $(x-9)(2x+7) =$ _____

9) $(5x+3)(x-4) =$ _____

10) $(4x+2)(3x-3) =$ _____

✎ **Factor each expression.**

11) $x^2 - 4x - 21 =$ _____

12) $x^2 + 14x + 45 =$ _____

13) $x^2 - 5x - 24 =$ _____

14) $x^2 - 7x + 6 =$ _____

15) $x^2 + 14x + 33 =$ _____

16) $4x^2 + 38x + 18 =$ _____

17) $5x^2 + 18x - 8 =$ _____

18) $2x^2 + 2x - 40 =$ _____

19) $2x^2 + 22x + 56 =$ _____

20) $12x^2 - 148x + 360 =$ _____

✎ **Calculate each equation.**

21) $(x+6)(x-9) = 0$

22) $(x+1)(x+11) = 0$

23) $(3x+9)(x+3) = 0$

24) $(5x-5)(6x+12) = 0$

25) $x^2 - 12x + 30 = 6$

26) $x^2 + 6x + 14 = 5$

27) $x^2 + \frac{9}{2}x + 7 = 5$

28) $x^2 + 2x - 25 = 10$

29) $2x^2 + 12x - 54 = 0$

30) $x^2 - 11x = 12$

Solving Quadratic Equations

✎ **Solve each equation by factoring or using the quadratic formula.**

1) $(x+5)(x-2) = 0$

2) $(x+8)(x+2) = 0$

3) $(x-9)(x+5) = 0$

4) $(x-3)(x-1) = 0$

5) $(x+9)(x+4) = 0$

6) $(2x+5)(x+9) = 0$

7) $(9x+8)(3x+9) = 0$

8) $(4x+2)(x+5) = 0$

9) $(x+2)(2x+9) = 0$

10) $(12x+3)(2x+9) = 0$

11) $2x^2 = 16x$

12) $x^2 - 16 = 0$

13) $2x^2 + 48 = 22x$

14) $-x^2 - 20 = 9x$

15) $x^2 + 8x = 33$

16) $2x^2 + 12x = 80$

17) $x^2 + 14x = -48$

18) $x^2 + 15x = -54$

19) $x^2 + 15x = -36$

20) $x^2 + 2x - 40 = 5x$

21) $x^2 + 16x = -63$

22) $x^2 - 18x = -81$

23) $10x^2 = 7x - 1$

24) $7x^2 - 5x + 8 = 8$

25) $8x^2 + 27 = 33x$

26) $5x^2 - 26x = -24$

27) $3x^2 + 6 = -19x$

28) $x^2 + 22x = -117$

29) $x^2 + 3x - 58 = 30$

30) $5x^2 + 20x - 200 = 25$

31) $3x^2 - 33x + 84 = 0$

32) $6x^2 - 31x + 30 = 15 - 10x^2$

Quadratic Formula and the Discriminant

✎ Find the value of the discriminant of each quadratic equation.

1) $3x(x - 9) = 0$

2) $2x^2 + 9x - 4 = 0$

3) $x^2 + 9x + 5 = 0$

4) $4x^2 - 4x + 7 = 0$

5) $x^2 + 7x - 6 = 0$

6) $4x^2 + 5x - 13 = 0$

7) $3x^2 + 7x + 11 = 0$

8) $x^2 - 4x - 12 = 0$

9) $5x^2 + 9x + 8 = 0$

10) $x^2 + 3x - 7 = 0$

11) $6x^2 + 7x - 13 = 0$

12) $-8x^2 - 11x + 9 = 0$

13) $-9x^2 - 13x + 7 = 0$

14) $-6x^2 - 7x - 9 = 0$

15) $14x^2 - 8x - 15 = 0$

16) $-9x^2 - 5x + 10 = 0$

17) $8x^2 + 9x - 14 = 0$

18) $7x^2 - 15x = 0$

19) $3x^2 - 7x + 9 = 0$

20) $7x^2 + 4x + 16 = 0$

✎ Find the discriminant of each quadratic equation then state the number of real and imaginary solutions.

21) $-x^2 - 4 = 4x$

22) $20x^2 = 20x - 5$

23) $-11x^2 - 11x = 22$

24) $19x^2 - 4x + 1 = 15x^2$

25) $-8x^2 = -6x + 6$

26) $2x^2 + 4x + 4 = 2$

27) $6x^2 - 2x - 9 = -12$

28) $-14x^2 - 56x - 64 = -8$

Quadratic Inequalities

✏️ **Solve each quadratic inequality.**

1) $x^2 - 64 < 0$

2) $-x^2 - 6x - 8 > 0$

3) $x^2 + 6x + 8 < 0$

4) $4x^2 + 28x + 40 > 0$

5) $5x^2 - 5x - 10 \geq 0$

6) $3x^2 > -12x - 27$

7) $4x^2 + 10x + 28 \leq 0$

8) $3x^2 - 9x - 30 \leq 0$

9) $5x^2 - 35x + 60 \geq 0$

10) $x^2 + 7x + 10 < 0$

11) $2x^2 + 16x - 130 > 0$

12) $8x^2 - 24x + 18 > 0$

13) $2x^2 - 32x + 136 \leq 0$

14) $x^2 - 14x + 49 \leq 0$

15) $2x^2 - 30x + 112 \geq 0$

16) $2x^2 + 16x + 32 \leq 0$

17) $x^2 - 121 \leq 0$

18) $2x^2 - 22x + 60 \geq 0$

19) $8x^2 + 10x + 18 \leq 0$

20) $4x^2 - 2x - 24 > 2x^2$

21) $4x^2 - 16x + 16 < 0$

22) $15x^2 - 6x \geq 14x^2 - 5$

23) $6x^2 - 24 > 4x^2 + 2x$

24) $3x^2 - x \geq 3x^2 - 4x + 6$

25) $2x^2 + 2x - 8 > x^2$

26) $4x^2 + 20x - 11 < 0$

27) $-2x^2 + 30x - 114 \geq 0$

28) $-8x^2 + 6x - 1 \leq 0$

29) $x^2 + 7x + 10 < 0$

30) $36x^2 + 46x + 10 \leq 0$

31) $5x^2 + 5x - 60 \geq 0$

32) $3x^2 + 4x \leq 2x^2 + 2x - 10$

Graphing Quadratic Functions

✎ **Sketch the graph of each function. Identify the vertex and axis of symmetry.**

1) $y = (x + 3)^2 + 5$

2) $y = (x - 3)^2 - 1$

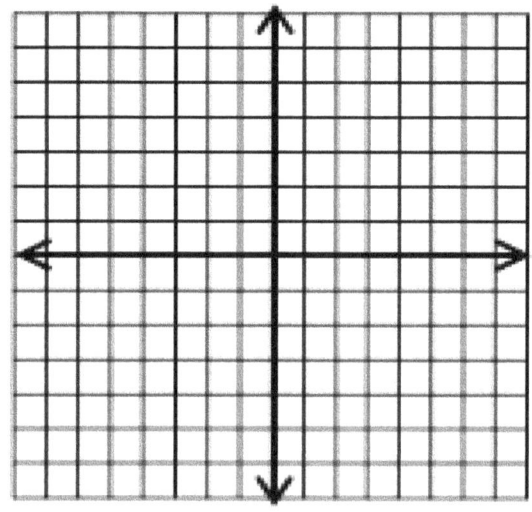
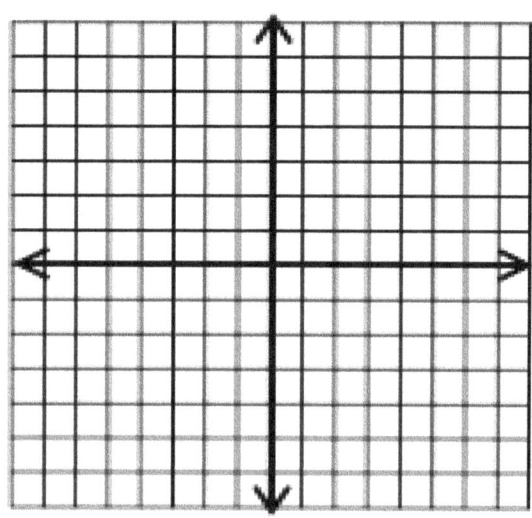

3) $y = 6 - (-x + 2)^2$

4) $y = -3x^2 - 6x + 9$

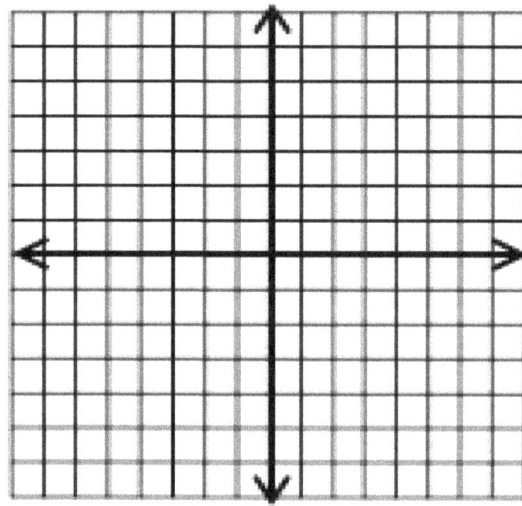
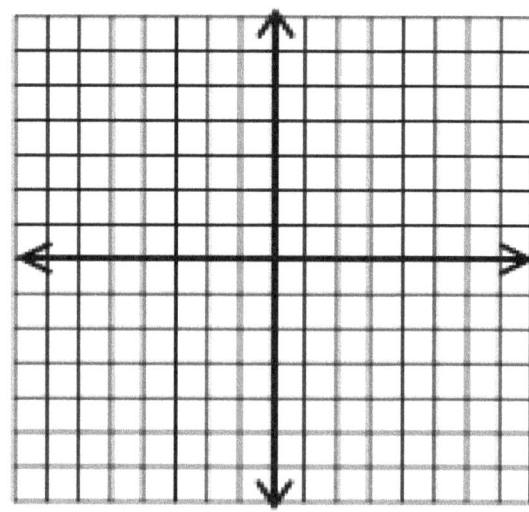

Domain and Range of Radical Functions

✏️ **Identify the domain and range of each function.**

1) $y = \sqrt{x+6} - 9$

2) $y = \sqrt[3]{5x-4} - 12$

3) $y = \sqrt{3x-9} + 7$

4) $y = \sqrt[3]{(8x+11)} - 9$

5) $y = 2\sqrt{6x+30} + 14$

6) $y = \sqrt[3]{(9x-15)} - 17$

7) $y = 2\sqrt{9x^2+18} + 7$

8) $y = \sqrt[3]{(8x^2-5)} - 13$

9) $y = \sqrt{2x^3+16} - 9$

10) $y = \sqrt[3]{(14x+3)} - 5x$

11) $y = 3\sqrt{-3(12x+24)} + 7$

12) $y = \sqrt[5]{(11x^2-17)} - 21$

13) $y = 4\sqrt{x-9} - 27$

14) $y = \sqrt[3]{5x+1} - 3$

✏️ **Sketch the graph of each function.**

15) $y = -3\sqrt{x} + 4$

16) $y = 6\sqrt{x} - 8$

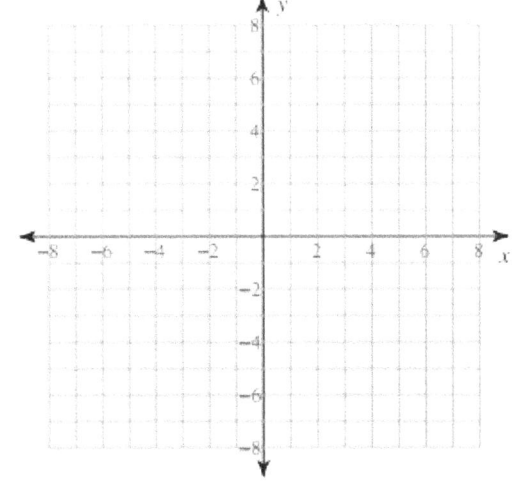

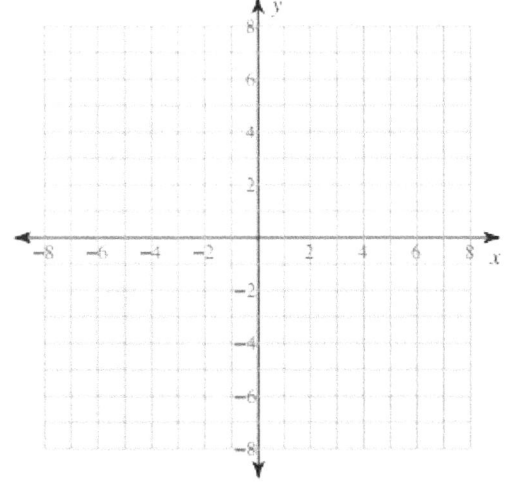

Solving Radical Equations

✎ Solve each equation. Remember to check for extraneous solutions.

1) $\sqrt{a} = 9$

2) $\sqrt{v} = 4$

3) $\sqrt{r} = 7$

4) $4 = 16\sqrt{x}$

5) $\sqrt{x+3} = 12$

6) $2 = \sqrt{x-8}$

7) $7 = \sqrt{r-6}$

8) $\sqrt{x-4} = 9$

9) $15 = \sqrt{x-6}$

10) $\sqrt{m+8} = 11$

11) $5\sqrt{3a} = 75$

12) $2\sqrt{10x} = 30$

13) $4 = \sqrt{3x-10}$

14) $\sqrt{150-3x} = 2$

15) $\sqrt{r+4} - 8 = 8$

16) $-18 = -3\sqrt{r+2}$

17) $60 = 6\sqrt{49v}$

18) $3 = \sqrt{50-x}$

19) $\sqrt{90-5a} = 6$

20) $\sqrt{-3n+33} = 3$

21) $\sqrt{21r-18} = 4r$

22) $\sqrt{-14+6x} = 7x$

23) $\sqrt{4x+15} = \sqrt{2x+11}$

24) $\sqrt{8v} = \sqrt{10v-14}$

25) $\sqrt{16-3x} = \sqrt{3x-8}$

26) $\sqrt{5m+12} = \sqrt{7m+12}$

27) $\sqrt{8r+15} = \sqrt{-13-5r}$

28) $\sqrt{4k+6} = \sqrt{2-8k}$

29) $-60\sqrt{x-16} = -120$

30) $\sqrt{20-x} = \sqrt{\dfrac{x}{4}}$

Answers of Worksheets – Chapter 6

Evaluating Function

1) $h(x) = -8x + 9$
2) $k(a) = 5a - 21$
3) $d(t) = 14t$
4) $y(x) = \frac{3}{17}x - \frac{9}{17}$
5) $m(n) = 18n - 94$
6) $c(p) = p^2 - 7p + 15$
7) -25
8) 6.5
9) -1
10) 14
11) -9
12) 38
13) 30
14) -29
15) -3
16) 11.5
17) 2
18) 3
19) 3
20) 12
21) $-\frac{109}{25}$
22) $-\frac{215}{27}$
23) 22
24) -308
25) 5
26) $-\frac{15b+7}{3b}$
27) $8a - 33$
28) $-2x + 7$
29) $2x^2 + 10$
30) $-16x^4 - 8$

Adding and Subtracting Functions

1) 1
2) 2
3) $4t - 6$
4) $-3x^2 - 5x - 7$
5) -26
6) 14
7) 2
8) $a^2 + a + 3$
9) $-8x^2 - 4x - 10$
10) $-z^2 - 4z - 4$
11) $4x^4 - x^2 + 9$
12) $-243x^6 + 12$

Multiplying and Dividing Functions

1) -90
2) -99
3) 180
4) 0
5) $\frac{81}{17}$
6) $\frac{3}{7}$
7) -154
8) -5
9) 0
10) $-\frac{1}{2}$
11) $x^4 - 7x^3 + 12x^2 + 9x - 36$
12) $3x^6 - 17x^4 + 20x^2$

Composition of Functions

1) -5
2) -1
3) 6
4) -2

5) 3
6) −10
7) $\frac{1}{16}$
8) 0
9) 16
10) $\frac{5}{8}$
11) 1
12) 1
13) 5
14) 42
15) −28
16) 2
17) 30
18) $25x - 8$
19) 3
20) −8
21) $3\sqrt{2}$
22) $\sqrt{11}$
23) −5
24) −10

Quadratic Equations

1) $x^2 + 6x - 16$
2) $x^2 + 10x + 9$
3) $x^2 + x - 30$
4) $x^2 + 4x - 21$
5) $x^2 - 17x + 72$
6) $4x^2 - 14x - 8$
7) $3x^2 + 6x - 24$
8) $2x^2 - 11x - 63$
9) $5x^2 - 17x - 12$
10) $12x^2 - 6x - 6$
11) $(x - 7)(x + 3)$
12) $(x + 5)(x + 9)$
13) $(x - 8)(x + 3)$
14) $(x - 1)(x - 6)$
15) $(x + 3)(x + 11)$
16) $(4x + 2)(x + 9)$
17) $(5x - 2)(x + 4)$
18) $(2x - 8)(x + 5)$
19) $(2x + 8)(x + 7)$
20) $4(x - 9)(3x - 10)$
21) $x = -6, x = 9$
22) $x = -1, x = -11$
23) $x = -3$
24) $x = 1, x = -2$
25) $x = 6$
26) $x = -3$
27) $x = -4, x = -\frac{1}{2}$
28) $x = 5, x = -7$
29) $x = 3, x = -9$
30) $x = -1, x = 12$

Solving quadratic equations

1) $\{-5, 2\}$
2) $\{-8, -2\}$
3) $\{9, -5\}$
4) $\{3, 1\}$
5) $\{-9, -4\}$
6) $\{-\frac{5}{2}, -9\}$
7) $\{-\frac{8}{9}, -3\}$
8) $\{-\frac{1}{2}, -5\}$
9) $\{-2, -\frac{9}{2}\}$
10) $\{-\frac{1}{4}, -\frac{9}{2}\}$
11) $\{8, 0\}$
12) $\{4, -4\}$
13) $\{3, 8\}$
14) $\{-5, -4\}$
15) $\{3, -11\}$
16) $\{4, -10\}$
17) $\{-6, -8\}$
18) $\{-6, -9\}$
19) $\{-3, -12\}$
20) $\{8, -5\}$
21) $\{-7, -9\}$
22) $\{9\}$
23) $\{\frac{1}{5}, \frac{1}{2}\}$
24) $\{\frac{5}{7}, 0\}$
25) $\{\frac{9}{8}, 3\}$
26) $\{\frac{6}{5}, 4\}$
27) $\{-\frac{1}{3}, -6\}$
28) $\{-9, -13\}$
29) $\{8, -11\}$
30) $\{5, -9\}$
31) $\{4, 7\}$
32) $\{\frac{15}{16}, 1\}$

Quadratic formula and the discriminant

1) 729
2) 113
3) 61
4) −96
5) 73
6) 233
7) 83
8) 8
9) −79
10) 37

11) 361
12) 409
13) 421
14) 7
15) 904
16) 385
17) 529
18) 225
19) −59
20) −432

21) 0, one real solution
22) 0, one real solution
23) −847, no solution
24) 0, one real solution
25) −156, no solution
26) 0, one real solution
27) −68, no solution
28) 0, one real solution

Solve quadratic inequalities

1) $-8 < x < 8$
2) $-4 < x < -2$
3) $-4 < x < -2$
4) $x < -5 \text{ or } x > -2$
5) $x \leq -1 \text{ or } x \geq 2$
6) all real numbers
7) no solution
8) $-2 \leq x \leq 5$
9) $x \leq 3 \text{ or } x \geq 4$
10) $-5 < x < -2$
11) $x < -13 \text{ or } x > 5$

12) $x < \frac{3}{2} \text{ or } x > \frac{3}{2}$
13) no solution
14) $x = 7$
15) $x \leq 7 \text{ or } x \geq 8$
16) $x = -4$
17) $-11 \leq x \leq 11$
18) $x \leq 5 \text{ or } x \geq 6$
19) no solution
20) $x < -3 \text{ or } x > 4$
21) no solution
22) $x \leq 1 \text{ or } x \geq 5$

23) $x < -3 \text{ or } x > 4$
24) $x \geq 2$
25) $x < -4 \text{ or } x > 2$
26) $-\frac{11}{2} < x < \frac{1}{2}$
27) no solution
28) $x \leq \frac{1}{4} \text{ or } x \geq \frac{1}{2}$
29) $-5 < x < -2$
30) $-1 \leq x \leq -\frac{5}{18}$
31) $x \leq -4 \text{ or } x \geq 3$
32) no solution

Graphing quadratic functions

1) $(-3, 5), x = -3$

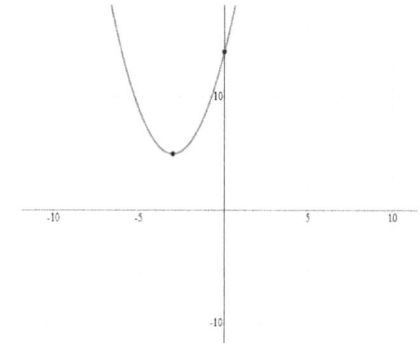

2) $(3, -1), x = 3$

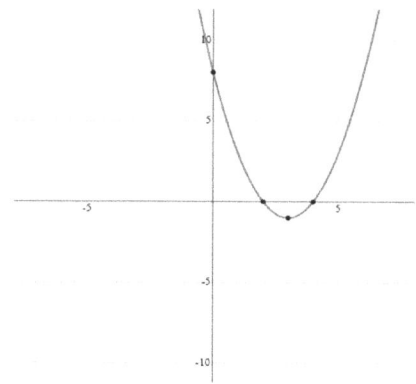

3) $(2, 6), x = 2$

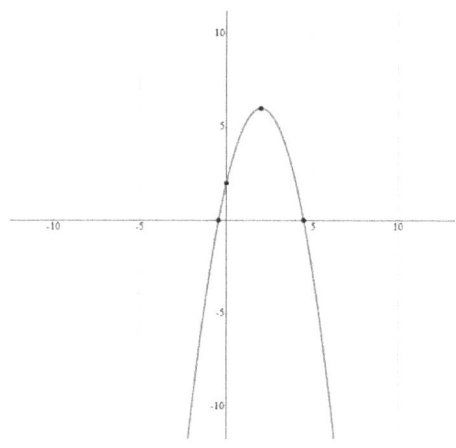

1) $(-1, 12), x = -1$

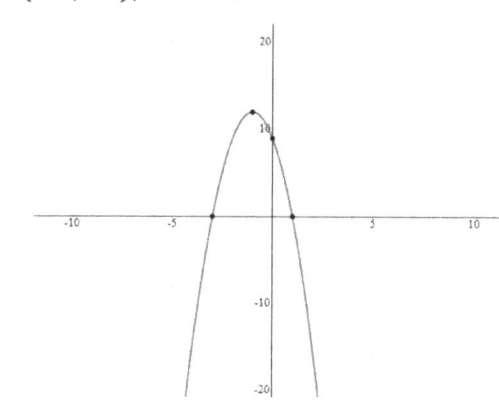

Domain and range of radical functions

1) domain: $x \geq -6$
   range: $y \geq -9$
2) domain: {all real numbers}
   range: {all real numbers}
3) domain: $x \geq 3$
   range: $y \geq 7$
4) domain: {all real numbers}
   range: {all real numbers}
5) domain: $x \geq -5$
   range: $y \geq 14$
6) domain: {all real numbers}
   range: {all real numbers}
7) domain: {all real numbers}
   range: $y \geq 6\sqrt{2} + 7$
8) domain: {all real numbers}
   range: {all real numbers}

9) domain: $x \geq -2$
   range: $y \geq -9$
10) domain: {all real numbers}
    range: {all real numbers}
11) domain: $x \leq -2$
    range: $y \geq 7$
12) domain: {all real numbers}
    range: {all real numbers}
13) domain: $x \geq 9$
    range: $y \geq -27$
14) domain: {all real numbers}
    range: {all real numbers}

15)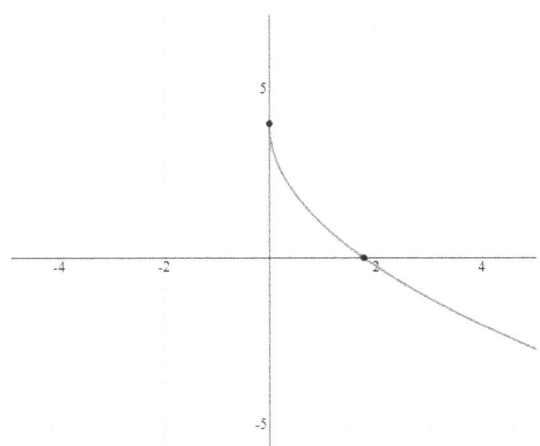

16)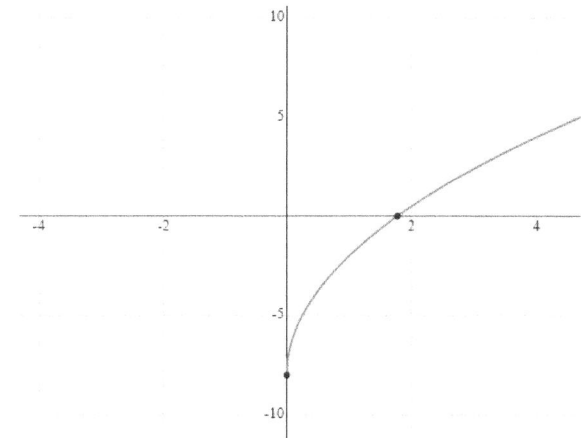

Solving radical equations

1) {81}
2) {16}
3) {49}
4) {$\frac{1}{16}$}
5) {141}
6) {12}
7) {55}
8) {85}
9) {231}
10) {113}
11) {75}
12) {22.5}
13) {$\frac{26}{3}$}
14) $\frac{146}{3}$
15) {252}
16) {34}
17) {$\frac{100}{49}$}
18) {41}
19) {54/5}
20) {8}
21) no solution
22) no solution
23) {−2}
24) {7}
25) {4}
26) {0}
27) no solution
28) $\left\{-\frac{1}{3}\right\}$
29) {20}
30) {16}

Chapter 7: Rational Expressions

Topics that you will learn in this chapter:

- ✓ Simplifying and Graphing Rational Expressions
- ✓ Adding and Subtracting Rational Expressions
- ✓ Multiplying and Dividing Rational Expressions
- ✓ Solving Rational Equations and Complex Fractions

"What music is to the heart, mathematics is to the mind."
— *Amit Kalantri*

Simplifying and Graphing Rational Expressions

✎ Simplify.

1) $\dfrac{x+2}{2x+4} =$

2) $\dfrac{2x^2+2x-12}{x+3} =$

3) $\dfrac{9}{3x-3} =$

4) $\dfrac{x^2-2x-3}{x^2+x-12} =$

5) $\dfrac{14x^3}{18x} =$

6) $\dfrac{x-2}{x^2+2x-8} =$

7) $\dfrac{x^2-6x-16}{x-8} =$

8) $\dfrac{25}{5x-5} =$

✎ Identify the points of discontinuity, holes, vertical asymptotes, x-intercepts, and horizontal asymptote of each.

9) $f(x) = \dfrac{x^3-x^2-6x}{-3x^3-3x+18} =$

10) $f(x) = \dfrac{x^2+x-6}{-4x^2-16x-12} =$

11) $f(x) = \dfrac{x-3}{x-9} =$

12) $f(x) = \dfrac{1}{2x^2-2x-12} =$

✎ Graph rational expressions.

13) $f(x) = \dfrac{x^2+2x-4}{x-2}$

14) $f(x) = \dfrac{4x^3-16x+64}{x^2-2x-4}$

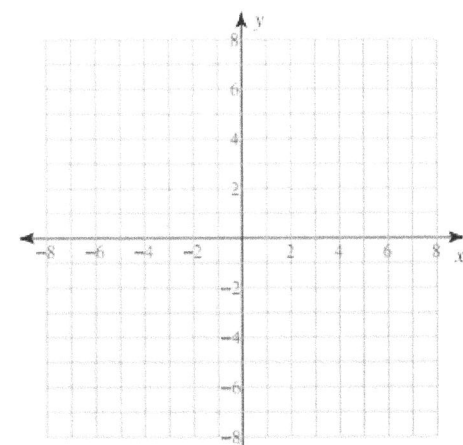
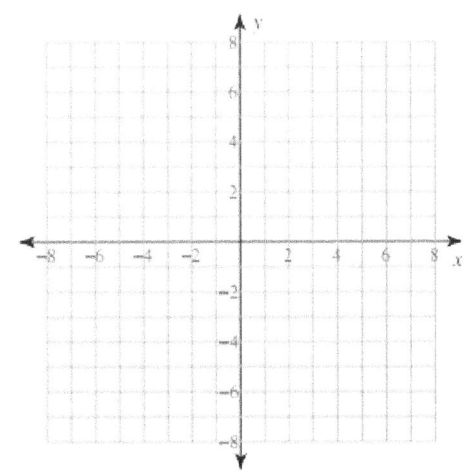

Adding and Subtracting Rational Expressions

✎ Simplify each expression.

1) $\dfrac{2}{4x+10} + \dfrac{x-4}{4x+10} =$

2) $\dfrac{x+3}{x-4} + \dfrac{x-3}{x+3} =$

3) $\dfrac{3}{x+6} - \dfrac{4}{x-7} =$

4) $\dfrac{x-4}{x^2-12} - \dfrac{x-1}{12-x^2} =$

5) $\dfrac{3}{x+3} + \dfrac{4x}{2x+6} =$

6) $\dfrac{5+x}{x} + \dfrac{x-2}{x} =$

7) $2 + \dfrac{x-2}{x+1} =$

8) $\dfrac{2x}{5x+4} + \dfrac{4x}{2x+3} =$

9) $\dfrac{3xy}{x^2-y^2} - \dfrac{x-y}{x+y} =$

10) $\dfrac{2}{x^2-5x+6} + \dfrac{-2}{x^2-9} =$

11) $\dfrac{4}{x+1} - \dfrac{2}{x+3} =$

12) $\dfrac{5x+5}{5x^2+15x-20} + \dfrac{3x}{2x} =$

13) $3 + \dfrac{x}{x+3} - \dfrac{3}{x^2-9} =$

14) $\dfrac{2}{x+1} - \dfrac{2}{x+3} =$

15) $\dfrac{2}{3x^2+18x} + \dfrac{4}{x} =$

16) $\dfrac{x^2+2x+1}{2x+2} + \dfrac{2x-2}{x-1} =$

17) $\dfrac{2x}{5x+4} + \dfrac{8x}{2x+3} =$

18) $\dfrac{x+4}{5+20x} - \dfrac{x-4}{5x^2+20x} =$

Multiplying and Dividing Rational Expressions

🖉 **Simplify each expression.**

1) $\dfrac{14x}{12} \times \dfrac{12}{16x} =$

2) $\dfrac{78x}{25} \times \dfrac{95}{21x^2} =$

3) $\dfrac{78}{3} \times \dfrac{44x}{93} =$

4) $\dfrac{65}{43} \times \dfrac{43x^2}{37} =$

5) $\dfrac{93}{21x} \times \dfrac{34x}{51x} =$

6) $\dfrac{5x + 20}{x + 4} \times \dfrac{x - 3}{5} =$

7) $\dfrac{x - 6}{x + 7} \times \dfrac{10x + 70}{x - 6} =$

8) $\dfrac{1}{x + 10} \times \dfrac{10x + 40}{x + 4} =$

9) $\dfrac{5(x+3)}{6x} \times \dfrac{7}{5(x+3)} =$

10) $\dfrac{9(x + 2)}{x + 2} \times \dfrac{9x}{9(x - 7)} =$

11) $\dfrac{3x^2 + 15x}{x + 5} \times \dfrac{1}{x + 7} =$

12) $\dfrac{15x^2 - 15x}{12x^2 - 12x} \times \dfrac{5x}{5x^2} =$

13) $\dfrac{1}{x - 8} \times \dfrac{x^2 + 5x - 24}{x + 8} =$

14) $\dfrac{x^2 - 10x + 25}{10x - 50} \times \dfrac{x - 5}{40 - 8x} =$

🖉 **Divide.**

15) $\dfrac{8 + 2x - x^2}{x^2 - 2x - 8} \div \dfrac{2x}{x + 6} =$

16) $\dfrac{8a}{a + 6} \div \dfrac{8a}{3a + 18} =$

17) $\dfrac{12x}{x - 5} \div \dfrac{12x}{12x - 60} =$

18) $\dfrac{x + 10}{7x^2 - 70x} \div \dfrac{1}{7x} =$

19) $\dfrac{x - 23}{x + 4x - 21} \div \dfrac{13x}{x + 8} =$

20) $\dfrac{4x}{x - 6} \div \dfrac{4x}{10x - 60} =$

21) $\dfrac{x + 5}{x + 12x + 35} \div \dfrac{4x}{x + 7} =$

22) $\dfrac{x + 3}{x + 13x + 40} \div \dfrac{6x}{x + 8} =$

23) $\dfrac{14x + 12}{2} \div \dfrac{63x + 54}{2x} =$

24) $\dfrac{8x^3 + 64x^2}{x^2 + 13x + 40} \div \dfrac{3(x+5)}{3x^3 + 15x^2} =$

25) $\dfrac{x^2 + 7x + 12}{x^2 + 5x + 6} \div \dfrac{1}{x + 2} =$

26) $\dfrac{x^2 - x - 6}{4x + 20} \div \dfrac{2}{x + 5} =$

27) $\dfrac{x + 4}{x^2 + 6x + 8} \div \dfrac{1}{x + 5} =$

28) $\dfrac{1}{3x} \div \dfrac{6x}{3x^2 + 21x} =$

Solving Rational Equations and Complex Fractions

✎ **Solve each equation. Remember to check for extraneous solutions.**

1) $\dfrac{2x-4}{x+1} = \dfrac{x+8}{x-2}$

2) $\dfrac{1}{x} = \dfrac{3}{4x} + 1$

3) $\dfrac{2x-3}{8x+1} = \dfrac{x+6}{x-2}$

4) $\dfrac{1}{4b^2} + \dfrac{1}{4b} = \dfrac{1}{b^2}$

5) $\dfrac{3x-2}{8x+1} = \dfrac{2x-5}{6x-5}$

6) $\dfrac{1}{n^2} - \dfrac{1}{n} = \dfrac{1}{2n^2}$

7) $\dfrac{1}{6b^2} = \dfrac{1}{3b^2} - \dfrac{1}{b}$

8) $\dfrac{1}{n-6} - 1 = \dfrac{5}{n-6}$

9) $\dfrac{5}{r-1} = -\dfrac{10}{r+1}$

10) $1 = \dfrac{1}{x^2+2x} + \dfrac{x+1}{x}$

11) $\dfrac{1}{x} = 5 + \dfrac{6}{9x}$

12) $\dfrac{x+6}{x^2-3x} - 1 = \dfrac{1}{x^2-3x}$

13) $\dfrac{x-3}{x+4} - 1 = \dfrac{1}{x+3}$

14) $\dfrac{1}{6x^2} = \dfrac{1}{2x^2} - \dfrac{1}{x}$

15) $\dfrac{x+5}{x^2+x} = \dfrac{1}{x^2+x} - \dfrac{x-6}{x+1}$

16) $1 = \dfrac{1}{x^2+2x} + \dfrac{x+1}{x}$

✎ **Simplify each expression.**

17) $\dfrac{\frac{2}{7}}{\frac{2}{35} - \frac{3}{16}} =$

18) $\dfrac{\frac{16}{3}}{-6\frac{2}{13}} =$

19) $\dfrac{8}{\frac{8}{x} + \frac{2}{3x}} =$

20) $\dfrac{x^2}{\frac{4}{6} - \frac{4}{x}} =$

21) $\dfrac{\frac{4}{x-3} - \frac{2}{x+2}}{\frac{6}{x^2+6x+8}} =$

22) $\dfrac{\frac{12}{x-1}}{\frac{12}{6} - \frac{12}{36}} =$

23) $\dfrac{2 + \frac{6}{x-3}}{2 - \frac{3}{x-3}} =$

24) $\dfrac{\frac{1}{2} - \frac{x+3}{4}}{\frac{x^2}{2} - \frac{3}{2}} =$

Answers of Worksheets – Chapter 7

Simplifying and Graphing rational expressions

1) $\frac{1}{2}$

2) $2(x-2)$

3) $\frac{3}{x-1}$

4) $\frac{x+1}{x+4}$

5) $\frac{7x^2}{2}$

6) $\frac{1}{x+4}$

7) $x+2$

8) $\frac{5}{x-1}$

9) Discontinuities: –3, 0; Vertical Asymptote: $x = -3, x = 2$; Holes: None
   Horizontal. Asymptote: None; x–intercepts: 0, –2, 3

10) Discontinuities –1, –3; Vertical Asymptote $x = -1$; Holes $x = -3$
    Horizontal Asymptote $y = -\frac{1}{4}$; x–intercepts. 2

11) Discontinuities: 9; Vertical Asymptote: $x = 9$; Holes: None
    Horizontal Asymptote: $y = 1$; x–intercepts: 2

12) Discontinuities: –2, 3; Vertical Asymptote: $x = -2, x = 3$; Holes: None
    Horizontal Asymptote: $y = 0$; x–intercepts: None

13)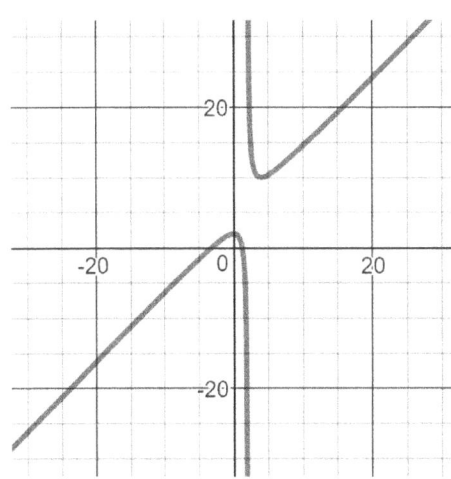

14)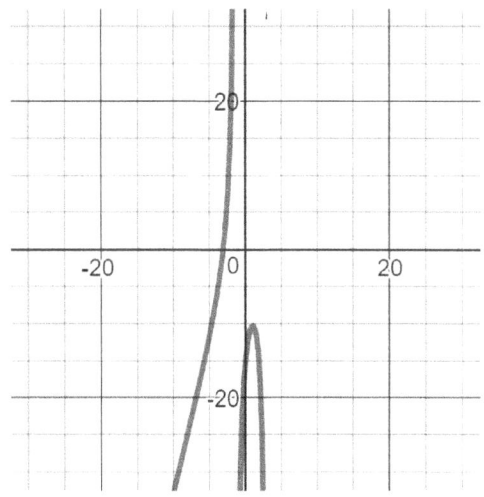

Adding and subtracting rational expressions

1) $\frac{-2+x}{4x+10}$

2) $\frac{2x^2 - x + 21}{(x-4)(x+3)}$

3) $\frac{-x-45}{(x+6)(x-7)}$

4) $\frac{2x-5}{x^2-12}$

5) $\frac{2x+3}{x+3}$

6) $2 + \frac{3}{x}$

7) $\frac{3x}{x+1}$

8) $\frac{24x^2 + 22x}{(5x+4)(2x+3)}$

9) $\frac{-x^2 + 5xy - y^2}{(x-y)(x+y)}$

10) $\dfrac{10}{(x^2-5x+6)(x+3)}$

11) $\dfrac{2x+10}{(x+1)(x+3)}$

12) $\dfrac{2x+2+3x^2}{3(3x-4+x^2)}$

13) $\dfrac{4x^2-3x-30}{(x+3)(x-3)}$

14) $\dfrac{4}{(x+1)(x+3)}$

15) $\dfrac{74+12x}{3x(x+6)}$

16) $\dfrac{x+1}{2}+2$

17) $\dfrac{44x^2+38x}{(5x+4)(2x+3)}$

18) $\dfrac{8}{5x^2+20x}$

Multiplying and Dividing rational expressions

1) $\dfrac{7}{8}$

2) $\dfrac{494}{35x}$

3) $\dfrac{1144x}{279}$

4) $\dfrac{65x^2}{37}$

5) $\dfrac{62}{21x}$

6) $x-3$

7) 10

8) $\dfrac{10}{x+10}$

9) $\dfrac{7}{6x}$

10) $\dfrac{9x}{x-7}$

11) $\dfrac{3x}{x+7}$

12) $\dfrac{5}{4x}$

13) $\dfrac{x-3}{x-8}$

14) $-\dfrac{(x-5)}{40}$

15) $\dfrac{-x}{2(x+6)}$

16) 3

17) 12

18) $\dfrac{x+10}{x-10}$

19) $\dfrac{x+8}{13x(x+7)}$

20) 10

21) $\dfrac{1}{4x}$

22) $\dfrac{x+3}{6x(x+5)}$

23) $\dfrac{2x}{9}$

24) $\dfrac{8x^4}{x+5}$

25) $x+4$

26) $\dfrac{(x-3)(x+2)}{8}$

27) $\dfrac{x+5}{x+2}$

28) $\dfrac{x+7}{6x}$

Solving rational equations and complex fractions

1) $\{0,17\}$

2) $\{\dfrac{1}{4}\}$

3) $\{\dfrac{-5}{2},-3\}$

4) $\{3\}$

5) $\{\dfrac{1}{6}\}$

6) $\{\dfrac{1}{2}\}$

7) $\{\dfrac{1}{6}\}$

8) $\{2\}$

9) $\{\dfrac{1}{3}\}$

10) $\{-3\}$

11) $\{\dfrac{1}{15}\}$

12) $\{5,-1\}$

13) $\{-\dfrac{25}{8}\}$

14) $\{\dfrac{1}{3}\}$

15) $\{1,4\}$

16) $\{-3\}$

17) $-\dfrac{160}{73}$

18) $-\dfrac{104}{93}$

19) $\dfrac{12x}{13}$

20) $\dfrac{3x^3}{2x-12}$

21) $\dfrac{(x+7)(x+4)}{3(x-3)}$

22) $\dfrac{36}{5x-5}$

23) $\dfrac{2x}{2x-9}$

24) $\dfrac{-1-x}{2x^2-6}$

Chapter 8:
Matrices

Topics that you will practice in this chapter:

- ✓ Adding and Subtracting Matrices
- ✓ Matrix Multiplications
- ✓ Finding Determinants of a Matrix
- ✓ Finding Inverse of a Matrix
- ✓ Matrix Equations

Mathematics is an independent world created out of pure intelligence.

− *William Woods Worth*

Adding and Subtracting Matrices

Simplify.

1) $\begin{vmatrix} -8 & 2 & -2 \end{vmatrix} + \begin{vmatrix} 0 & -2 & -5 \end{vmatrix}$

2) $\begin{vmatrix} 3 & 1 \\ -2 & -1 \\ -5 & 0 \end{vmatrix} + \begin{vmatrix} 2 & -2 \\ 2 & 1 \\ 5 & 3 \end{vmatrix}$

3) $\begin{vmatrix} -1 & 1 & -3 \\ 5 & -3 & -2 \end{vmatrix} - \begin{vmatrix} 7 & -1 & -3 \\ 2 & 3 & -5 \end{vmatrix}$

4) $\begin{vmatrix} 8 & 1 \end{vmatrix} + \begin{vmatrix} -5 & -6 \end{vmatrix}$

5) $\begin{vmatrix} 2 \\ 3 \end{vmatrix} + \begin{vmatrix} 2 \\ 7 \end{vmatrix}$

6) $\begin{vmatrix} -3r + 2t \\ -3r \\ 2s \end{vmatrix} + \begin{vmatrix} 3r \\ -t \\ -5r + 3 \end{vmatrix}$

7) $\begin{vmatrix} 2z - 1 \\ -6 \\ -3 - 6z \\ 5y \end{vmatrix} + \begin{vmatrix} -3y \\ 5z \\ 7 + 2z \\ 6z \end{vmatrix}$

8) $\begin{vmatrix} -3n & n + 7m \\ -n & -5m \end{vmatrix} + \begin{vmatrix} 2 & -3 \\ 3m & 0 \end{vmatrix}$

9) $\begin{vmatrix} 4 & 6 \\ -7 & 2 \end{vmatrix} - \begin{vmatrix} 1 & -2 \\ 3 & 9 \end{vmatrix}$

10) $\begin{vmatrix} 4 & -7 & 6 \\ 3 & -5 & 5 \\ -5 & 5 & -10 \end{vmatrix} + \begin{vmatrix} 0 & 6 & -2 \\ 6 & 4 & 6 \\ 3 & -7 & -4 \end{vmatrix}$

Matrix Multiplication

Simplify.

1) $\begin{vmatrix} -2 & -2 \\ -3 & 1 \end{vmatrix} \times \begin{vmatrix} -1 & -2 \\ 2 & 3 \end{vmatrix}$

2) $\begin{vmatrix} 3 & 2 \\ -1 & 0 \\ -2 & 3 \end{vmatrix} \times \begin{vmatrix} -1 & 4 \\ -1 & -3 \end{vmatrix}$

3) $\begin{vmatrix} 3 & 1 & 0 \\ 2 & 5 & 4 \end{vmatrix} \times \begin{vmatrix} 2 & 5 & 1 \\ 1 & -2 & 1 \end{vmatrix}$

4) $\begin{vmatrix} -5 \\ 1 \\ 3 \end{vmatrix} \times \begin{vmatrix} 3 & -4 \end{vmatrix}$

5) $\begin{vmatrix} 5 & -2 \\ 1 & 1 \\ 0 & -4 \end{vmatrix} \times \begin{vmatrix} -2 & 2 \\ 1 & 0 \end{vmatrix}$

6) $\begin{vmatrix} 1 & 1 \\ -3 & 0 \end{vmatrix} \cdot \begin{vmatrix} 5 & -1 \\ 1 & 0 \end{vmatrix}$

7) $\begin{vmatrix} -3 & -2y \\ x & -1 \end{vmatrix} \cdot \begin{vmatrix} -2x & 1 \\ -y & -1 \end{vmatrix}$

8) $\begin{vmatrix} 2 & -3v \end{vmatrix} \cdot \begin{vmatrix} -u & -2v \\ 1 & 0 \end{vmatrix}$

9) $\begin{vmatrix} -2 & 4 & 0 \\ 0 & 2 & -1 \\ 3 & -2 & 3 \\ -3 & 4 & 1 \end{vmatrix} \cdot \begin{vmatrix} 4 & 0 \\ 3 & -3 \\ 2 & 1 \end{vmatrix}$

10) $\begin{vmatrix} 2 & 1 & 0 \\ 0 & 2 & 1 \end{vmatrix} \cdot \begin{vmatrix} -1 & 3 \\ -1 & 4 \\ 3 & -2 \end{vmatrix}$

11) $\begin{vmatrix} -3 & 4 \\ -2 & 3 \end{vmatrix} \cdot \begin{vmatrix} 0 & -2 \\ 3 & 1 \end{vmatrix}$

12) $\begin{vmatrix} 1 & 0 \\ -5 & -2 \end{vmatrix} \cdot \begin{vmatrix} 2 & -2 \\ 1 & 1 \end{vmatrix}$

13) $\begin{vmatrix} 1 & 3 \\ -1 & -2 \end{vmatrix} \cdot \begin{vmatrix} 1 & -4 \\ 0 & 4 \end{vmatrix}$

14) $\begin{vmatrix} -2 & -2 \\ 1 & 0 \\ 2 & 0 \\ 2 & -1 \end{vmatrix} \times \begin{vmatrix} 0 & -1 & 2 \\ -1 & 0 & -4 \end{vmatrix}$

Finding Determinants of a Matrix

Evaluate the determinant of each matrix.

1) $\begin{vmatrix} 4 & 0 \\ -9 & -5 \end{vmatrix}$

2) $\begin{vmatrix} 9 & 5 \\ 1 & 0 \end{vmatrix}$

3) $\begin{vmatrix} -2 & 2 \\ 3 & 3 \end{vmatrix}$

4) $\begin{vmatrix} -1 & 7 \\ -2 & 8 \end{vmatrix}$

5) $\begin{vmatrix} -5 & 2 \\ 2 & 3 \end{vmatrix}$

6) $\begin{vmatrix} 7 & -4 \\ 0 & 8 \end{vmatrix}$

7) $\begin{vmatrix} 1 & -3 \\ 8 & -5 \end{vmatrix}$

8) $\begin{vmatrix} 3 & 5 \\ 4 & 1 \end{vmatrix}$

9) $\begin{vmatrix} 5 & 4 \\ -2 & 8 \end{vmatrix}$

10) $\begin{vmatrix} 3 & 2 \\ 0 & 3 \end{vmatrix}$

11) $\begin{vmatrix} 4 & -1 & 2 \\ 0 & 1 & -2 \\ 2 & 3 & 3 \end{vmatrix}$

12) $\begin{vmatrix} -2 & 1 & -2 \\ -2 & 3 & 1 \\ 1 & 0 & 3 \end{vmatrix}$

13) $\begin{vmatrix} 4 & 2 & 2 \\ 2 & -1 & 3 \\ 0 & 5 & 4 \end{vmatrix}$

14) $\begin{vmatrix} 2 & -1 & 0 \\ 3 & 2 & -2 \\ 2 & 1 & 2 \end{vmatrix}$

15) $\begin{vmatrix} 3 & 1 & 0 \\ 0 & -2 & -2 \\ 1 & 5 & 4 \end{vmatrix}$

Finding Inverse of a Matrix

✎Find the inverse of each matrix.

1) $\begin{vmatrix} 3 & 5 \\ 1 & 4 \end{vmatrix}$

2) $\begin{vmatrix} 3 & 2 \\ 4 & 3 \end{vmatrix}$

3) $\begin{vmatrix} 5 & 4 \\ 2 & 2 \end{vmatrix}$

4) $\begin{vmatrix} 6 & 5 \\ 2 & 4 \end{vmatrix}$

5) $\begin{vmatrix} -4 & 3 \\ 2 & 4 \end{vmatrix}$

6) $\begin{vmatrix} 5 & 2 \\ 7 & 6 \end{vmatrix}$

7) $\begin{vmatrix} 1 & 0 \\ 9 & 4 \end{vmatrix}$

8) $\begin{vmatrix} -8 & -9 \\ 3 & 4 \end{vmatrix}$

9) $\begin{vmatrix} -2 & 7 \\ -2 & 7 \end{vmatrix}$

10) $\begin{vmatrix} -3 & 2 \\ 5 & 4 \end{vmatrix}$

11) $\begin{vmatrix} 8 & 4 \\ 1 & 2 \end{vmatrix}$

12) $\begin{vmatrix} 1 & 8 \\ 2 & 0 \end{vmatrix}$

13) $\begin{vmatrix} 1 & 9 \\ 0 & 0 \end{vmatrix}$

14) $\begin{vmatrix} 10 & 6 \\ 5 & 3 \end{vmatrix}$

Matrix Equations

✏️ *Solve each equation.*

1) $\begin{vmatrix} -2 & 4 \\ 0 & 1 \end{vmatrix} z = \begin{vmatrix} 8 \\ 5 \end{vmatrix}$

2) $3x = \begin{vmatrix} 15 & -6 \\ 9 & -12 \end{vmatrix}$

3) $\begin{vmatrix} -4 & 3 \\ -9 & 5 \end{vmatrix} = \begin{vmatrix} 1 & 6 \\ 3 & 7 \end{vmatrix} - x$

4) $Y - \begin{vmatrix} -3 \\ -5 \\ 11 \\ 11 \end{vmatrix} = \begin{vmatrix} -2 \\ 8 \\ -18 \\ -2 \end{vmatrix}$

5) $\begin{vmatrix} -2 & -1 \\ 1 & -3 \end{vmatrix} C = \begin{vmatrix} 5 \\ -6 \end{vmatrix}$

6) $\begin{vmatrix} -1 & -2 \\ 4 & 3 \end{vmatrix} B = \begin{vmatrix} 0 & -1 & -1 \\ -5 & 14 & -1 \end{vmatrix}$

7) $\begin{vmatrix} -2 & 2 \\ 3 & -1 \end{vmatrix} C = \begin{vmatrix} 10 \\ -9 \end{vmatrix}$

8) $\begin{vmatrix} 2 & 5 \\ 1 & 1 \end{vmatrix} C = \begin{vmatrix} 4 \\ -1 \end{vmatrix}$

9) $\begin{vmatrix} 0 & -5 \\ 2 & 4 \end{vmatrix} Z = \begin{vmatrix} 15 \\ 0 \end{vmatrix}$

10) $\begin{vmatrix} -9 \\ 6 \\ -15 \end{vmatrix} = 3B$

11) $\begin{vmatrix} -7 \\ 4 \\ 2 \end{vmatrix} = y - \begin{vmatrix} 5 \\ -4 \\ -8 \end{vmatrix}$

12) $-3b - \begin{vmatrix} 8 \\ 4 \\ -2 \end{vmatrix} = \begin{vmatrix} -26 \\ -7 \\ -16 \end{vmatrix}$

Answers of Worksheets – Chapter 8

Adding and Subtracting Matrices

1) $|-8 \quad 0 \quad -7|$

2) $\begin{vmatrix} 5 & -1 \\ 0 & 0 \\ 0 & 3 \end{vmatrix}$

3) $\begin{vmatrix} -8 & 2 & 0 \\ 3 & -6 & 3 \end{vmatrix}$

4) $|3 \quad -5|$

5) $\begin{vmatrix} 4 \\ 10 \end{vmatrix}$

6) $\begin{vmatrix} 2t \\ -3r - t \\ 2s - 5r + 3 \end{vmatrix}$

7) $\begin{vmatrix} 2z - 1 - 3y \\ -6 + 5z \\ 4 - 4z \\ 5y + 6z \end{vmatrix}$

8) $\begin{vmatrix} -3n + 2 & n + 7m - 3 \\ -n + 3m & -5m \end{vmatrix}$

9) $\begin{vmatrix} 3 & 8 \\ -10 & -7 \end{vmatrix}$

10) $\begin{vmatrix} 4 & -1 & 4 \\ 9 & -1 & 11 \\ -2 & -2 & -14 \end{vmatrix}$

Matrix Multiplication

1) $\begin{vmatrix} -2 & -2 \\ 5 & 9 \end{vmatrix}$

2) $\begin{vmatrix} -5 & 6 \\ 1 & -4 \\ -1 & -17 \end{vmatrix}$

3) Undefined

4) $\begin{vmatrix} -15 & 20 \\ 3 & -4 \\ 9 & -12 \end{vmatrix}$

5) $\begin{vmatrix} -12 & 10 \\ -1 & 2 \\ -4 & 0 \end{vmatrix}$

6) $\begin{vmatrix} 6 & -1 \\ -15 & 3 \end{vmatrix}$

7) $\begin{vmatrix} 6x + 2y^2 & 2y - 3 \\ -2x^2 + y & x + 1 \end{vmatrix}$

8) $|-2u - 3v \quad -4v|$

9) $\begin{vmatrix} 4 & -12 \\ 4 & -7 \\ 12 & 9 \\ 2 & -11 \end{vmatrix}$

10) $\begin{vmatrix} -3 & 10 \\ 1 & 6 \end{vmatrix}$

11) $\begin{vmatrix} 12 & 10 \\ 9 & 7 \end{vmatrix}$

12) $\begin{vmatrix} 2 & -2 \\ -12 & 8 \end{vmatrix}$

13) $\begin{vmatrix} 1 & 8 \\ -1 & -4 \end{vmatrix}$

14) $\begin{vmatrix} 2 & 2 & 4 \\ 0 & -1 & 2 \\ 0 & -2 & 4 \\ 1 & -2 & 8 \end{vmatrix}$

Finding Determinants of a Matrix

1) −20

2) −5

3) −12

4) 6

5) −19

6) 56

7) 19

8) −17

9) 48

10) 9

11) 36

12) −5

13) −72

14) 22

15) 4

Finding Inverse of a Matrix

1) $\begin{vmatrix} \frac{4}{7} & -\frac{5}{7} \\ \frac{-1}{7} & \frac{3}{7} \end{vmatrix}$

2) $\begin{vmatrix} 3 & -2 \\ -4 & 3 \end{vmatrix}$

3) $\begin{vmatrix} 1 & -2 \\ -1 & \frac{5}{2} \end{vmatrix}$

4) $\begin{vmatrix} \frac{2}{7} & \frac{-5}{14} \\ \frac{-1}{7} & \frac{3}{7} \end{vmatrix}$

5) $\begin{vmatrix} -\frac{2}{11} & \frac{3}{22} \\ \frac{1}{11} & \frac{2}{11} \end{vmatrix}$

6) $\begin{vmatrix} \frac{3}{8} & -\frac{1}{8} \\ -\frac{5}{16} & \frac{5}{16} \end{vmatrix}$

7) $\begin{vmatrix} 1 & 0 \\ -\frac{9}{4} & \frac{1}{4} \end{vmatrix}$

8) $\begin{vmatrix} -\frac{4}{5} & -\frac{9}{5} \\ \frac{3}{5} & \frac{8}{5} \end{vmatrix}$

9) No inverse exists

10) $\begin{vmatrix} -\frac{2}{11} & \frac{1}{11} \\ \frac{5}{22} & \frac{3}{22} \end{vmatrix}$

11) $\begin{vmatrix} \frac{1}{6} & -\frac{1}{3} \\ -\frac{1}{12} & \frac{2}{3} \end{vmatrix}$

12) $\begin{vmatrix} 0 & \frac{1}{2} \\ \frac{1}{8} & -\frac{1}{16} \end{vmatrix}$

13) No inverse exists

14) No inverse exists

Matrix Equations

1) $\begin{vmatrix} 6 \\ 5 \end{vmatrix}$

2) $\begin{vmatrix} 5 & -2 \\ 3 & -4 \end{vmatrix}$

3) $\begin{vmatrix} 5 & 3 \\ 12 & 2 \end{vmatrix}$

4) $\begin{vmatrix} -5 \\ 3 \\ -7 \\ 9 \end{vmatrix}$

5) $\begin{vmatrix} -3 \\ 1 \end{vmatrix}$

6) $\begin{vmatrix} -2 & 5 & -1 \\ 1 & -2 & 1 \end{vmatrix}$

7) $\begin{vmatrix} -2 \\ 3 \end{vmatrix}$

8) $\begin{vmatrix} -3 \\ 2 \end{vmatrix}$

9) $\begin{vmatrix} 6 \\ -3 \end{vmatrix}$

10) $\begin{vmatrix} -3 \\ 2 \\ -5 \end{vmatrix}$

11) $\begin{vmatrix} -2 \\ 0 \\ -6 \end{vmatrix}$

12) $\begin{vmatrix} 6 \\ 1 \\ 6 \end{vmatrix}$

Chapter 9:

Sequences and Series

Topics that you will practice in this chapter:

- ✓ Arithmetic Sequences
- ✓ Geometric Sequences
- ✓ Comparing Arithmetic and Geometric Sequences
- ✓ Finite Geometric Series
- ✓ Infinite Geometric Series

Mathematics is like checkers in being suitable for the young, not too difficult, amusing, and without peril to the state. — Plato

Arithmetic Sequences

✎ **Find the next three terms of each arithmetic sequence.**

1) $32, 23, 14, 5, -4, \ldots$

2) $-91, -63, -35, -7, \ldots$

3) $51, 62, 73, 84, 95, \ldots$

4) $84, 53, 22, -9, -40, \ldots$

✎ **Given the first term and the common difference of an arithmetic sequence find the first five terms and the explicit formula.**

5) $a_1 = 9, d = 12$

6) $a_1 = -10, d = -5$

7) $a_1 = 52, d = 22$

8) $a_1 = 210, d = -102$

✎ **Given a term in an arithmetic sequence and the common difference find the first five terms and the explicit formula.**

9) $a_{51} = -468, d = -12$

10) $a_{31} = 230, d = 6$

11) $a_{62} = -128.2, d = -4.2$

12) $a_{33} = -2{,}352, d = -77$

✎ **Given a term in an arithmetic sequence and the common difference find the recursive formula and the three terms in the sequence after the last one given.**

13) $a_{25} = -156, d = -6$

14) $a_{16} = 111, d = 7.1$

15) $a_{22} = 43, d = 1.8$

16) $a_{14} = -17, d = 0.4$

Geometric Sequences

✎ **Determine if the sequence is geometric. If it is, find the common ratio.**

1) $2, -14, 98, -686, \ldots$

2) $-3, -15, -75, -375, \ldots$

3) $7, 17, 31, 126, \ldots$

4) $-5, -35, -245, -1715, \ldots$

✎ **Given the first term and the common ratio of a geometric sequence find the first five terms and the explicit formula.**

5) $a_1 = 0.7, r = -3$

6) $a_1 = 0.4, r = 5$

✎ **Given the recursive formula for a geometric sequence find the common ratio, the first five terms, and the explicit formula.**

7) $a_n = a_{n-1} \times 6, a_1 = 2$

8) $a_n = a_{n-1} \cdot (-4), a_1 = -6$

9) $a_n = a_{n-1} \cdot 9, a_1 = 0.2$

10) $a_n = a_{n-1} \cdot 3, a_1 = -8$

✎ **Given two terms in a geometric sequence find the 9th term and the recursive formula.**

11) $a_5 = 729$ and $a_6 = -243$

12) $a_6 = -768$ and $a_3 = 12$

Comparing Arithmetic and Geometric Sequences

✎ **For each sequence, state if it is arithmetic, geometric, or neither.**

1) $5, 10, 15, 20, \ldots$

2) $6, 10, 14, 20, \ldots$

3) $2, 6, 24, 51, \ldots$

4) $1, 8, 18, 28, 36, \ldots$

5) $2, 8, 17, 52, 142, \ldots$

6) $2, 5, 9, 17, 36, \ldots$

7) $0.6, 3, 15, 75, 375, \ldots$

8) $4, 20, 100, 500, \ldots$

9) $-18, -23, -28, -33, -38, \ldots$

10) $-3, 12, -48, 192, -768, \ldots$

11) $8, 18, 26, 39, 50, \ldots$

12) $3, 12, 90, 150, 210 \ldots$

13) $-22, -12, -2, 2, 12, \ldots$

14) $a_n = 2 \cdot 7^{n-1}$

15) $a_n = 8 \cdot 4^{n-1}$

16) $a_n = 9 - 5n$

17) $a_n = -110 + 200n$

18) $a_n = 15 + 13n$

19) $a_n = -6 \cdot (-11)^{n-1}$

20) $a_n = 120 + 42n$

21) $a_n = (4n)^4$

22) $a_n = 28 + 6n$

23) $a_n = -(13)^{n-1}$

24) $a_n = -7 \cdot (1.5)^{n-1}$

25) $a_n = \frac{2n+1}{7n}$

26) $a_n = \frac{24-13n}{6n}$

27) $a_n = \frac{8-17n}{2n}$

28) $a_n = \frac{32 - a_{n-1}}{9}$

29) $a_n = -\frac{3}{19} + \frac{2}{7}n$

Finite Geometric Series

✎ **Evaluate the related series of each sequence.**

1) $-2, 8, -32, 128$

2) $-1, 3, -9, 27, -81$

3) $-1, 4, -16, 64, -256$

4) $1, 8, 64, 512$

5) $-6, -24, -96, -384$

6) $2, -12, 72, -432, 2592$

✎ **Evaluate each geometric series described.**

7) $1 + 3 + 9 + 27 \ldots, n = 6$ _____

8) $1.5 - 6 + 24 - 96 \ldots, n = 6$ _____

9) $-2 - 6 - 18 - 54 \ldots, n = 7$ _____

10) $0.5 - 3 + 18 - 108 \ldots, n = 6$ _____

11) $2.5 - 10 + 40 - 160 \ldots, n = 8$ _____

12) $-1 + 7 - 49 + 343 \ldots, n = 6$ _____

13) $a_1 = -2, r = 6, n = 5$ _____

14) $a_1 = 3, r = 2, n = 9$ _____

15) $\sum_{n=1}^{5} 4 \cdot (-3)^{n-1}$ _____

16) $\sum_{n=1}^{7} 6 \cdot (-2)^{n-1}$ _____

17) $\sum_{n=1}^{5} 3 \cdot (5)^{n-1}$ _____

18) $\sum_{m=1}^{10} (-2)^{m-1}$ _____

19) $\sum_{m=1}^{4} 8 \times (5)^{m-1}$ _____

20) $\sum_{k=1}^{8} 2 \times (4)^{k-1}$ _____

Infinite Geometric Series

✎ Determine if each geometric series converges or diverges.

1) $a_1 = -1.4, \ r = 6$

2) $a_1 = 5.2, r = 0.3$

3) $a_1 = -6, r = 7.2$

4) $a_1 = 12, r = 0.04$

5) $a_1 = 3, r = 15$

6) $-1, 7, -49, 343, ...$

7) $6, -1, \frac{1}{6}, -\frac{1}{36}, \frac{1}{216}, \ ...$

8) $512 + 64 + 8 + 1 \ ...$

9) $-4 + \frac{12}{7} - \frac{36}{49} + \frac{108}{343} \ ...$

10) $\frac{120}{459} - \frac{60}{153} + \frac{30}{51} - \frac{15}{17} \ ...$

✎ Evaluate each infinite geometric series described.

11) $a_1 = 4, r = -\frac{1}{6}$

12) $a_1 = 18, r = -\frac{1}{3}$

13) $a_1 = 16, r = -\frac{1}{7}$

14) $a_1 = 8, r = \frac{1}{3}$

15) $2 + 0.5 + 0.125 + 0.031 + \cdots$

16) $125 - 25 + 5 - 1 \ ...,$

17) $1 - 0.6 + 0.36 - 0.216 \ ...,$

18) $3 + \frac{12}{5} + \frac{48}{25} + \frac{192}{125} \ ...,$

19) $\sum_{k=1}^{\infty} 11^{k-1}$

20) $\sum_{i=1}^{\infty} \left(\frac{2}{5}\right)^{i-1}$

21) $\sum_{k=1}^{\infty} \left(-\frac{3}{7}\right)^{k-1}$

22) $\sum_{n=1}^{\infty} 12\left(\frac{5}{6}\right)^{n-1}$

Answers of Worksheets – Chapter 9

Arithmetic Sequences

1) $-13, -22, -31$
2) $21, 49, 77$
3) $106, 117, 128$
4) $-71, -102, -133$
5) First Five Terms: $9, 21, 33, 45, 57$, Explicit: $a_n = 9 + 12(n-1)$
6) First Five Terms: $-10, -15, -20, -25, -30$, Explicit: $a_n = -10 - 5(n-1)$
7) First Five Terms: $52, 74, 96, 118, 140$, Explicit: $a_n = 52 + 22(n-1)$
8) First Five Terms: $210, 108, 6, -96, -198$, Explicit: $a_n = 210 - 102(n-1)$
9) First Five Terms: $132, 120, 108, 96, 84$, Explicit: $a_n = 132 - 12(n-1)$
10) First Five Terms: $50, 56, 62, 68, 74$, Explicit: $a_n = 50 + 6(n-1)$
11) First Five Terms: $128, 123.8, 119.6, 115.4, 111.2$, Explicit: $a_n = 128 - 4.2(n-1)$
12) First Five Terms: $112, 35, -42, -119, -196$, Explicit: $a_n = 112 - 77(n-1)$
13) Next 3 terms: $-162, -168, -174$, Recursive: $a_n = a_{n-1} - 6, a_1 = -6$
14) Next 3 terms: $118.2, 125.2, 132.3$ Recursive: $a_n = a_{n-1} + 7.1, \ a_1 = 4.5$
15) Next 3 terms: $44.8, 46.6, 48.4$, Recursive: $a_n = a_{n-1} + 1.8, a_1 = 5.2$
16) Next 3 terms: $-9.8, -9.6, -9.4$, Recursive: $a_n = a_{n-1} + 0.4, a_1 = -22.2$

Geometric Sequences

1) $r = -7$
2) $r = 5$
3) not geometric
4) $r = 7$
5) First Five Terms: $0.7, -2.1, 6.3, -18.9, 56.7$
   Explicit: $a_n = 0.7 \times (-3)^{n-1}$
6) First Five Terms: $0.4, 2, 10, 50, 250$
   Explicit: $a_n = 0.4 \times (5)^{n-1}$
7) Common Ratio: $r = 6$
   First Five Terms: $2, 12, 72, 432, 2{,}592$
   Explicit: $a_n = 2 . (6)^{n-1}$

8) Common Ratio: $r = -4$

First Five Terms: $-6, 24, -96, 384, -1{,}536$

Explicit: $a_n = -6 \cdot (-4)^{n-1}$

9) Common Ratio: $r = 9$

First Five Terms: $0.2;\ 1.8;\ 16.2;\ 145.8;\ 1{,}312.2;\ 11{,}809.8$

Explicit: $a_n = 0.2 \cdot (9)^{n-1}$

10) Common Ratio: $r = 3$

First Five Terms: $-8, -24, -72, -216, -648$

Explicit: $a_n = -8 \cdot (3)^{n-1}$

11) $a_9 = 9$, Recursive: $a_n = a_{n-1} \cdot (\frac{-1}{3})$, $a_1 = 59{,}049$

12) $a_9 = 49{,}152$, Recursive: $a_n = a_{n-1} \cdot (-4)$, $a_1 = 0.75$

Comparing Arithmetic and Geometric Sequences

1) Arithmetic
2) Arithmetic
3) Neither
4) Neither
5) Neither
6) Neither
7) Geometric
8) Geometric
9) Arithmetic
10) Geometric
11) Neither
12) Neither
13) Arithmetic
14) Geometric
15) Geometric
16) Arithmetic
17) Arithmetic
18) Arithmetic
19) Geometric
20) Arithmetic
21) Neither
22) Arithmetic
23) Geometric
24) Geometric
25) Neither
26) Neither
27) Neither
28) Neither
29) Arithmetic

Finite Geometric

1) 102
2) −61
3) −205
4) 585
5) −510
6) 2,157
7) 364
8) −1,228.5
9) −2,186
10) −3,333.5
11) −32,767.5
12) 14,699
13) −3,110
14) 1,533
15) 244
16) 258
17) 2,343
18) −341
19) 1,248
20) 43,680

Infinite Geometric

1) Diverges
2) Converges
3) Diverges
4) Converges
5) Diverges
6) Diverges
7) Converges
8) Converges
9) Converges
10) Diverges
11) $\frac{24}{7}$
12) $\frac{27}{2}$
13) 14
14) 12
15) $\frac{8}{3}$
16) $\frac{625}{6}$
17) $\frac{5}{8}$
18) 15
19) Infinite
20) $\frac{5}{3}$
21) $\frac{7}{10}$
22) 72

Chapter 10:
Complex Numbers

Topics that you will practice in this chapter:

- ✓ Adding and Subtracting Complex Numbers
- ✓ Multiplying and Dividing Complex Numbers
- ✓ Graphing Complex Numbers
- ✓ Rationalizing Imaginary Denominators

Mathematics is a hard thing to love. It has the unfortunate habit, like a rude dog, of turning its most unfavorable side towards you when you first make contact with it. — David Whiteland

Adding and Subtracting Complex Numbers

✎ **Simplify.**

1) $(8i) - (4i) =$

2) $(5i) + (2i) =$

3) $(2i) + (7i) =$

4) $(-6i) - (i) =$

5) $(12i) + (4i) =$

6) $(4i) - (-4i) =$

7) $(-4i) + (-5i) =$

8) $(13i) - (6i) =$

9) $(-21i) - (7i) =$

10) $(-4i) + (2 + 8i) =$

11) $(8 - 4i) + (-6i) =$

12) $(-3i) + (9 + 12i) =$

13) $5 + (9 - 2i) =$

14) $(10i) - (-6 + 2i) =$

15) $(3 + 9i) - (-4i) =$

16) $(7 + 8i) + (-5i) =$

17) $(5i) - (-3i + 4) =$

18) $(6i + 2) + (-2i) =$

19) $(12) - (18 + 3i) =$

20) $(7 + 3i) + (6 + 2i) =$

21) $(4 - 9i) + (3 + 8i) =$

22) $(7 + 3i) + (10 + 12i) =$

23) $(-5 + 5i) - (-5 - 7i) =$

24) $(-8 + 12i) - (-9 + 8i) =$

25) $(-18 + 3i) - (-3 - 12i) =$

26) $(-13 - 4i) + (9 + 12i) =$

27) $(-15 - 2i) - (-14 - 6i) =$

28) $-4 + (8i) + (-14 + 7i) =$

29) $19 - (8i) + (2 - 5i) =$

30) $-3 + (-4 - 8i) - 9 =$

31) $(-24i) + (5 - 8i) + 12 =$

32) $(-11i) - (15 - 12i) + 9i =$

Multiplying and Dividing Complex Numbers

✎ **Simplify.**

1) $(5i)(-3i) =$

2) $(-7i)(2i) =$

3) $(3i)(3i)(-3i) =$

4) $(6i)(-6i) =$

5) $(-7 - 6i)(7 + 6i) =$

6) $(4 - 2i)^2 =$

7) $(5 - 2i)(4 - 2i) =$

8) $(5 + 2i)^2 =$

9) $(7i)(-3i)(9 - 2i) =$

10) $(2 - 8i)(6 - 8i) =$

11) $(-9 + 3i)(1 + 4i) =$

12) $(7 - 8i)(9 - 3i) =$

13) $5(3i) - (5i)(-4 + 2i) =$

14) $\dfrac{5}{-25i} =$

15) $\dfrac{12-9i}{-3i} =$

16) $\dfrac{4+9i}{i} =$

17) $\dfrac{20i}{-6+2i} =$

18) $\dfrac{-4-11i}{2i} =$

19) $\dfrac{7i}{3-i} =$

20) $\dfrac{4-9i}{12-5i} =$

21) $\dfrac{8-3i}{-4-4i} =$

22) $\dfrac{-9-5i}{-3-i} =$

23) $\dfrac{-4+i}{-6-5i} =$

24) $\dfrac{-6-7i}{-3+4i} =$

25) $\dfrac{8+11i}{5-5i} =$

Graphing Complex Numbers

✎ **Identify each complex number graphed.**

1)

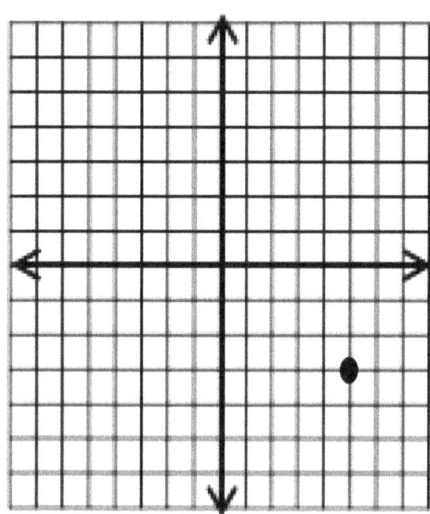

2)

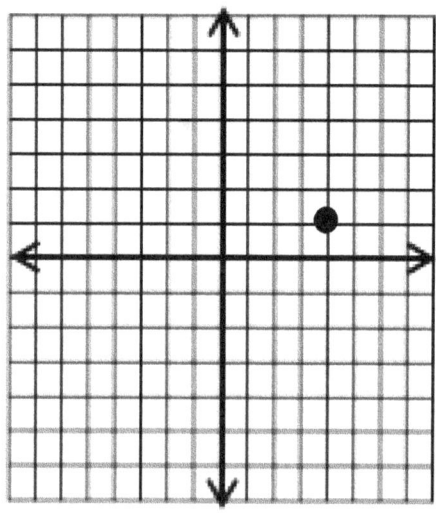

3)

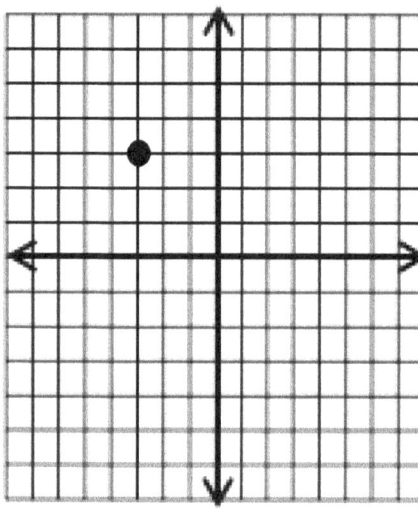

4)

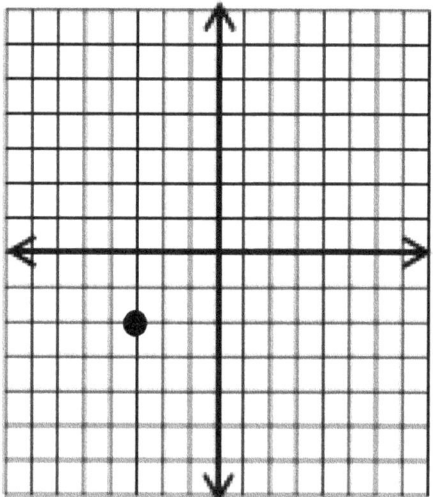

Rationalizing Imaginary Denominators

✎ **Simplify.**

1) $\dfrac{-8}{-8i} =$

2) $\dfrac{-3}{-21i} =$

3) $\dfrac{-14}{-28i} =$

4) $\dfrac{-30}{-5i} =$

5) $\dfrac{9}{2i} =$

6) $\dfrac{27}{-9i} =$

7) $\dfrac{45}{-20i} =$

8) $\dfrac{-26}{8i} =$

9) $\dfrac{6x}{3yi} =$

10) $\dfrac{9-9i}{-3i} =$

11) $\dfrac{4-9i}{-i} =$

12) $\dfrac{12+4i}{3i} =$

13) $\dfrac{8i}{-1+4i} =$

14) $\dfrac{8i}{-2+6i} =$

15) $\dfrac{-15-3i}{-4+4i} =$

16) $\dfrac{-5-9i}{3+4i} =$

17) $\dfrac{-11-4i}{5-2i} =$

18) $\dfrac{-4+6i}{-3i} =$

19) $\dfrac{9+5i}{4i} =$

20) $\dfrac{-5-3i}{7-2i} =$

21) $\dfrac{-8+i}{-3i} =$

22) $\dfrac{9+i}{-5-2i} =$

23) $\dfrac{-9-5i}{-7-2i} =$

24) $\dfrac{9i-5}{-3-6i} =$

Answers of Worksheets – Chapter 11

Adding and Subtracting Complex Numbers

1) $4i$
2) $7i$
3) $9i$
4) $-7i$
5) $16i$
6) $8i$
7) $-9i$
8) $7i$
9) $-28i$
10) $2 + 4i$
11) $8 - 10i$
12) $9 + 9i$
13) $14 - 2i$
14) $6 + 8i$
15) $3 + 13i$
16) $7 + 3i$
17) $-4 + 8i$
18) $2 + 4i$
19) $-6 - 3i$
20) $13 + 5i$
21) $7 - i$
22) $17 + 15i$
23) $12i$
24) $1 + 4i$
25) $-15 + 15i$
26) $-4 + 8i$
27) $-1 + 4i$
28) $-18 + 15i$
29) $21 - 13i$
30) $-16 - 8i$
31) $17 - 32i$
32) $-15 + 10i$

Multiplying and Dividing Complex Numbers

1) 15
2) 14
3) $27i$
4) 36
5) $-13 - 84i$
6) $-16i + 12$
7) $16 - 18i$
8) $21 + 20i$
9) $189 - 42i$
10) $-52 - 64i$
11) $-21 - 33i$
12) $39 - 93i$
13) $10 + 35i$
14) $\frac{i}{5}$
15) $3 + 4i$
16) $9 - 4i$
17) $1 - 3i$
18) $\frac{11}{2} - 2i$
19) $-\frac{7}{10} + \frac{21}{10}i$
20) $\frac{93}{169} - \frac{88}{169}i$
21) $-\frac{5}{8} + \frac{11}{8}i$
22) $\frac{16}{5} + \frac{3}{5}i$
23) $\frac{19}{61} - \frac{26}{61}i$
24) $-\frac{2}{5} + \frac{9}{5}i$
25) $-\frac{3}{10} + \frac{19}{10}i$

Graphing Complex Numbers

2) $5 - 3i$
3) $4 + i$
4) $-3 + 3i$
5) $-3 - 2i$

Rationalizing Imaginary Denominators

1) $-i$
2) $-\frac{1}{7}i$
3) $\frac{-1}{2}i$
4) $-6i$
5) $-\frac{9}{2}i$
6) $3i$
7) $\frac{9}{4}i$
8) $\frac{13}{4}i$
9) $-\frac{2x}{y}i$
10) $3 + 3i$
11) $9 + 4i$
12) $-\frac{4}{3} + 4i$
13) $\frac{32}{17} - \frac{8}{17}i$
14) $\frac{6}{5} - \frac{2}{5}i$
15) $\frac{3}{2} + \frac{9}{4}i$
16) $-\frac{51}{25} - \frac{7}{25}i$
17) $-\frac{47}{29} - \frac{42}{29}i$
18) $-2 - \frac{4}{3}i$
19) $-\frac{5}{4} + \frac{9}{4}i$
20) $-\frac{29}{53} - \frac{31}{53}i$
21) $-\frac{1}{3} - \frac{8}{3}i$
22) $-\frac{47}{29} + \frac{13}{29}i$
23) $\frac{73}{53} + \frac{17}{53}i$
24) $-\frac{13}{15} - \frac{19}{15}i$

Chapter 11: Logarithms

Topics that you will practice in this chapter:

- ✓ Rewriting Logarithms
- ✓ Evaluating Logarithms
- ✓ Properties of Logarithms
- ✓ Natural Logarithms
- ✓ Exponential Equations Requiring Logarithms
- ✓ Solving Logarithmic Equations

Mathematics is an art of human understanding. — William Thurston

Rewriting Logarithms

✎ **Rewrite each equation in exponential form.**

1) $\log_5 25 = 2$

2) $\log_4 256 = 4$

3) $\log_3 81 = 4$

4) $\log_8 64 = 2$

5) $\log_6 216 = 3$

6) $\log_2 16 = 4$

7) $\log_{10} 100 = 2$

8) $\log_3 243 = 5$

9) $\log_5 625 = 4$

10) $\log_2 256 = 8$

11) $\log_3 6,561 = 8$

12) $\log_{11} 121 = 2$

13) $\log_{14} 196 = 2$

14) $\log_{81} 3 = \frac{1}{4}$

15) $\log_{27} 3 = \frac{1}{3}$

16) $\log_{32} 2 = \frac{1}{5}$

17) $\log_{512} 8 = \frac{1}{3}$

18) $\log_2 \frac{1}{8} = -3$

19) $\log_2 \frac{1}{16} = -4$

20) $\log_a \frac{7}{3} = b$

✎ **Rewrite each exponential equation in logarithmic form.**

21) $12^2 = 144$

22) $7^3 = 343$

23) $4^5 = 1,024$

24) $15^2 = 225$

25) $5^4 = 625$

26) $6^4 = 1,296$

27) $2^9 = 512$

28) $5^5 = 3,125$

29) $4^{-6} = \frac{1}{4,096}$

30) $3^{-5} = \frac{1}{243}$

31) $16^{-2} = \frac{1}{256}$

32) $6^{-3} = \frac{1}{216}$

33) $3^{-9} = \frac{1}{19,683}$

34) $21^{-2} = \frac{1}{441}$

Evaluating Logarithms

✎ **Evaluate each logarithm.**

1) $\log_3 2{,}187 =$

2) $\log_2 256 =$

3) $\log_5 125 =$

4) $\log_5 625 =$

5) $\log_3 243 =$

6) $\log_4 1{,}024 =$

7) $\log_8 64 =$

8) $\log_8 \frac{1}{8} =$

9) $\log_6 \frac{1}{36} =$

10) $\log_2 \frac{1}{16} =$

11) $\log_6 \frac{1}{216} =$

12) $\log_3 \frac{1}{256} =$

13) $\log_{18} \frac{1}{324} =$

14) $\log_{256} \frac{1}{4} =$

15) $\log_{512} 8 =$

16) $\log_4 \frac{1}{4{,}096} =$

17) $\log_9 \frac{1}{729} =$

18) $\log_{216} \frac{1}{6} =$

✎ **Circle the points which are on the graph of the given logarithmic functions.**

19) $y = 5\log_8(3x - 4) + 1$ (6, 5),   (4, 6),   (4, 8)

20) $y = 3\log_2(4x) - 6$ (4, 6), $(\frac{1}{4}, 16)$, $(\frac{1}{4}, -6)$

21) $y = -3\log_5(x - 2) + 5$ (7, -2),   (7, 2),   (6, -3)

22) $y = \frac{1}{2}\log_6(6x) + 4$ (6, 5), $(6, \frac{1}{5})$,   (6, -5)

23) $y = -\log_9 9(x + 5) + 4$ (-4, 2),   (4, 0),   (4, 2)

24) $y = -\log_8(x - 6) - 4$ $(7, -\frac{1}{4})$,   (7, -4), $(7, -\frac{1}{4})$

25) $y = -3\log_6(x + 3) + 6$ (3, 3),   (-5, -3),   (-5, 3)

Properties of Logarithms

✎ **Expand each logarithm.**

1) $\log(9 \times 4) =$

2) $\log(6 \times 3) =$

3) $\log(2 \times 8) =$

4) $\log\left(\frac{8}{7}\right) =$

5) $\log\left(\frac{9}{5}\right) =$

6) $\log\left(\frac{4}{11}\right)^3 =$

7) $\log(9 \times 4^3) =$

8) $\log\left(\frac{7}{3}\right)^2 =$

9) $\log\left(\frac{5^4}{9}\right) =$

10) $\log(x \times y)^7 =$

11) $\log(x^2 \times y \times z^5) =$

12) $\log\left(\frac{u^8}{v}\right) =$

13) $\log\left(\frac{x}{y^4}\right) =$

✎ **Condense each expression to a single logarithm.**

14) $\log 7 - \log 12 =$

15) $\log 8 + \log 3 =$

16) $4\log 2 - 7\log 5 =$

17) $6\log 4 - 9\log 5 =$

18) $3\log 8 - \log 17 =$

19) $8\log 3 - 6\log 2 =$

20) $\log 11 - 2\log 5 =$

21) $4\log 6 + 3\log 9 =$

22) $4\log 5 + 5\log 13 =$

23) $7\log_5 a + 16\log_5 b =$

24) $5\log_6 x - 7\log_6 y =$

25) $\log_5 u - 9\log_5 v =$

26) $8\log_3 u + 21\log_3 v =$

27) $26\log_7 u - 15\log_7 v =$

Natural Logarithms

✎ **Solve each equation for** x.

1) $e^x = 9$

2) $e^x = 36$

3) $e^x = 49$

4) $\ln x = 3$

5) $\ln(\ln x) = 3$

6) $e^x = 11$

7) $\ln(5x + 9) = 1$

8) $\ln(7x + 3) = 3$

9) $\ln(8x + 5) = 4$

10) $\ln x = \frac{1}{3}$

11) $\ln 6x = e^4$

12) $\ln x = \ln 4 + \ln 7$

13) $\ln x = 3\ln 3 + \ln 8$

✎ **Evaluate without using a calculator.**

14) $3\ln e =$

15) $\ln e^{10} =$

16) $4 \ln e =$

17) $\ln e^{15} =$

18) $13\ln e =$

19) $3\ln e^4 =$

20) $e^{\ln 19} =$

21) $e^{2\ln 5} =$

22) $e^{4\ln 3} =$

23) $\ln \sqrt[6]{e} =$

✎ **Reduce the following expressions to simplest form.**

24) $e^{-2\ln 6 + 2\ln 4} =$

25) $e^{-2\ln\left(\frac{4}{5e}\right)} =$

26) $2\ln(e^6) =$

27) $\ln\left(\frac{1}{e}\right)^9 =$

28) $e^{\ln 6 + 3\ln 5} =$

29) $e^{\ln\left(\frac{13}{e}\right)} =$

30) $7\ln(1^{-3e}) =$

31) $\ln\left(\frac{1}{e}\right)^{-12} =$

32) $3\ln\left(\frac{\sqrt[6]{e}}{3e}\right) =$

33) $e^{-3\ln e + 3\ln 3} =$

34) $e^{\ln\frac{15}{e}} =$

35) $19\ln(e^e) =$

Exponential Equations and Logarithms

 Solve each equation for the unknown variable.

1) $3^{2n} = 27$
2) $5^r = 125$
3) $15^n = 85$
4) $8^{r+3} = 2$
5) $144^x = 12$
6) $7^{-3v-2} = 49$
7) $8^{2n} = 64$
8) $6^n = 1{,}296$
9) $\frac{15^{2a}}{3^{-a}} = 315$
10) $11 \times 11^{-v} = 1{,}331$
11) $3^{2n} = \frac{1}{81}$
12) $(\frac{1}{9})^n = 81$
13) $256^{2x} = 4$
14) $9^{3-2x} = 9^{-x}$
15) $6^{-3x} = 6^{x-3}$
16) $2^{3n} = 32$
17) $12^{5x+3} = 12^{2x}$
18) $10^{2n} = 100$
19) $3^{-4k} = 243$
20) $3^r = 9^{-4r}$
21) $13^{x+3} = 13^{4x}$
22) $9^{3x} = 729$
23) $15 \times 15^{-v} = 225$
24) $\frac{81}{3^{-2m}} = 3^{-2m-1}$
25) $8^{-2n} \times 8^2 = 8^{-n}$
26) $(\frac{1}{9})^{2n+1} \times (\frac{1}{9})^{-n-10} = (\frac{1}{9})^{-2n}$

✎ **Solve each problem. (Round to the nearest whole number)**

27) A substance decays 15% each day. After 11 days, there are 6 milligrams of the substance remaining. How many milligrams were there initially? _____

28) A culture of bacteria grows continuously. The culture doubles every 4 hours. If the initial number of bacteria is 13, how many bacteria will there be in 23 hours? _____

29) Bob plans to invest $12,000 at an annual rate of 6.5%. How much will Bob have in the account after six years if the balance is compounded quarterly? _____

30) Suppose you plan to invest $8,000 at an annual rate of 6%. How much will you have in the account after 4 years if the balance is compounded monthly? _____

Solving Logarithmic Equations

✎ **Find the value of the variables in each equation.**

1) $\log(x) + 8 = 4$

2) $-\log_3 4x = 5$

3) $\log(x) + 7 = 6$

4) $\log x - \log 7 = 4$

5) $\log x + \log 4 = 2$

6) $\log 4 + \log x = 3$

7) $\log x + \log 2 = \log 12$

8) $-3\log_3(x - 2) = -15$

9) $\log 4x = \log(3x + 2)$

10) $\log(2k - 4) = \log(k - 5)$

11) $\log(5p - 2) = \log(-2p + 12)$

12) $-8 + \log_3(n + 3) = -8$

13) $\log_3(x + 5) = \log_3(x^2 + 8)$

14) $\log_9(v^2 + 24) = \log_9(-3v - 8)$

15) $\log(9 + 4b) = \log(7b^2 + 6b)$

16) $\log_9(x + 8) - \log_9 x = \log_9 7$

17) $\log_5 9 + \log_5 x^2 = \log_5 81$

18) $\log_6(x + 5) + \log_6 x = \log_6 24$

✎ **Find the value of x in each natural logarithm equation.**

19) $\ln 9 - \ln(3x + 9) = 3$

20) $\ln(x - 4) - \ln(x - 3) = \ln 4$

21) $\ln e^{27} - \ln(x + 3) = 3$

22) $\ln(2x - 6) - \ln(x - 12) = \ln 10$

23) $\ln 6x + \ln(x - 2) = \ln 3x$

24) $\ln(x - 3) - 2\ln(x - 3) = \ln 9$

25) $\ln(9x + 3) - \ln 5 = 6$

26) $\ln(x - 5) + \ln(x - 4) = \ln 2$

27) $\ln 8 + \ln(x + 4) = 10$

28)

29) $3\ln 3x - \ln(x + 9) = 3\ln 3x$

30) $\ln x^2 + \ln x^4 = \ln 1$

31) $\ln x^6 - \ln(x + 6) = 6\ln x$

32) $16\ln(x - 2) = 4\ln(x^2 - 4x + 4)$

33) $\ln(x^2 + 10) = \ln(3x + 8)$

34) $6\ln x - 6\ln(x + 3) = 12\ln(x^2)$

35) $\ln(2x - 3) - \ln(4x - 3) = \ln 4$

36) $\ln 3 + 9\ln(x + 2) = \ln 3$

37) $3\ln e^2 + \ln(3x - 2) = \ln 3 + 9$

Answers of Worksheets – Chapter 11

Rewriting Logarithms

1) $5^2 = 25$
2) $4^4 = 256$
3) $3^4 = 81$
4) $8^2 = 64$
5) $6^3 = 216$
6) $2^4 = 16$
7) $10^2 = 100$
8) $3^5 = 243$
9) $5^4 = 625$
10) $2^8 = 256$
11) $3^8 = 6,561$
12) $11^2 = 121$
13) $14^2 = 196$
14) $81^{\frac{1}{4}} = 3$
15) $27^{\frac{1}{3}} = 3$
16) $32^{\frac{1}{5}} = 2$
17) $512^{\frac{1}{3}} = 8$
18) $2^{-3} = \frac{1}{8}$
19) $2^{-4} = \frac{1}{16}$
20) $a^b = \frac{7}{3}$
21) $\log_{12} 144 = 2$
22) $\log_3 343 = 7$
23) $\log_4 1,024 = 5$
24) $\log_{15} 225 = 2$
25) $\log_5 625 = 4$
26) $\log_6 1,296 = 4$
27) $\log_2 512 = 9$
28) $\log_5 3,125 = 5$
29) $\log_4 \frac{1}{4,096} = -6$
30) $\log_3 \frac{1}{243} = -5$
31) $\log_{16} \frac{1}{256} = -2$
32) $\log_6 \frac{1}{216} = -3$
33) $\log_3 \frac{1}{19,683} = -9$
34) $\log_{21} \frac{1}{441} = -2$

Evaluating Logarithms

1) 7
2) 8
3) 3
4) 4
5) 5
6) 5
7) 2
8) -1
9) -2
10) -4
11) -3
12) -4
13) -2
14) $-\frac{1}{4}$
15) $\frac{1}{3}$
16) -6
17) -3
18) $-\frac{1}{3}$
19) $(4, 6)$
20) $(\frac{1}{4}, -6)$
21) $(7, 2)$
22) $(6, 5)$
23) $(4, 2)$
24) $(7, -4)$
25) $(3, 3)$

Properties of Logarithms

1) $\log 9 + \log 4$
2) $\log 6 + \log 3$
3) $\log 2 + \log 8$
4) $\log 8 - \log 7$
5) $\log 9 - \log 5$
6) $3 \log 4 - 3 \log 11$
7) $\log 9 + 3 \log 4$
8) $2\log 7 - 2 \log 3$

9) $4\log 5 - \log 9$

10) $7\log x + 7\log y$

11) $2\log x + \log y + 5\log z$

12) $8\log u - \log v$

13) $\log x - 4\log y$

14) $\log \frac{7}{12}$

15) $\log(8 \times 3)$

16) $\log \frac{2^4}{5^7}$

17) $\log \frac{4^6}{9^5}$

18) $\log \frac{8^3}{17}$

19) $\log \frac{3^8}{2^6}$

20) $\log \frac{11}{5^2}$

21) $\log(6^4 \times 9^3)$

22) $\log(5^4 \times 13^5)$

23) $\log_5(a^7 b^{16})$

24) $\log_6 \frac{x^5}{y^7}$

25) $\log_5 \frac{u}{v^9}$

26) $\log_3(u^8 \times v^{21})$

27) $\log_7 \frac{u^{26}}{v^{15}}$

Natural Logarithms

1) $x = \ln 9$
2) $x = \ln 36, x = 2\ln(6)$
3) $x = \ln 49, x = 2\ln(7)$
4) $x = e^3$
5) $x = e^{e^3}$
6) $x = \ln 11$
7) $x = \frac{e-9}{5}$
8) $x = \frac{e^3 - 3}{7}$
9) $x = \frac{e^4 - 5}{8}$
10) $x = \sqrt[3]{e}$
11) $x = \frac{e^{e^4}}{6}$
12) $x = 28$
13) $x = 216$
14) 3
15) 10
16) 4
17) 15
18) 13
19) 12
20) 19
21) 25
22) 81
23) $\frac{1}{6}$
24) $\frac{4}{9}$
25) $\frac{25}{16e^2}$
26) 12
27) -9
28) 750
29) $\frac{13}{e}$
30) 0
31) 12
32) -5.8
33) $27e^{-3} = \frac{27}{e^3}$
34) $\frac{15}{e}$
35) $19e$

Exponential Equations and Logarithms

1) $\frac{3}{2}$
2) 3
3) 1.64
4) $\frac{-8}{3}$
5) $\frac{1}{2}$
6) $-\frac{4}{3}$
7) 1
8) 0.883

9) -2
10) -2
11) -2
12) $\frac{1}{8}$
13) 3
14) $\frac{3}{4}$
15) $\frac{5}{3}$

16) -1
17) 1
18) $-\frac{5}{4}$
19) 0
20) 1
21) 1
22) -1
23) -1.25

24) 2
25) 3
26) 35.9
27) 699.6
28) $\$17,668.3$
29) $\$10,163.9$

Solving Logarithmic Equations

1) $\{\frac{1}{10,000}\}$
2) $\{\frac{1}{972}\}$
3) $\{\frac{1}{10}\}$
4) $\{70,000\}$
5) $\{25\}$
6) $\{250\}$
7) $\{6\}$
8) $\{245\}$
9) $\{2\}$
10) No Solution
11) $\{2\}$
12) $\{-2\}$
13) No Solution

14) No Solution
15) $\{1, -\frac{9}{7}\}$
16) $\{\frac{4}{3}\}$
17) $\{3, -3\}$
18) $\{3\}$
19) $x = \frac{3-3e^3}{e^3}$
20) No Solution
21) $e^{24} - 3$
22) $\{\frac{57}{4}\}$
23) $\{\frac{5}{2}\}$
24) $\{\frac{28}{9}\}$

25) $x = \frac{5e^6-3}{9}$
26) $x = 6$
27) $x = \frac{e^{10}-32}{8}$
28) No Solution
29) $\{1, -1\}$
30) No Solution
31) $x = 3$
32) $\{1, 2\}$
33) $\{0.64951 ...\}$
34) No Solution
35) $\{-1\}$
36) $x = \frac{3e^3+2}{3}$

Chapter 12: Conic Sections

Topics that you'll learn in this chapter:

- ✓ Equation of a Parabola
- ✓ Focus, Vertex, and Directrix of a Parabola
- ✓ Standard Form of a Circle
- ✓ Standard Equation of an Ellipse
- ✓ Hyperbola in Standard Form
- ✓ Conic Sections in Standard Form

Equation of a Parabola

✎ **Write the equation of the following parabolas.**

1) Vertex (0, 0) and Focus (0, 2)

2) Vertex (3, 2) and Focus (3, 4)

3) Vertex (1, 1) and Focus (1, 6)

4) Vertex (– 1, 2) and Focus (– 1, 5)

5) Vertex (2, 2) and Focus (2, 6)

6) Vertex (0, 1) and Focus (0, 2)

7) Vertex (2, 1) and Focus (4, 1)

8) Vertex (5, 0) and Focus (9, 0)

9) Vertex (– 2, 4) and Focus (2, 4)

10) Vertex (– 4, 2) and Focus (0, 2)

Focus, Vertex, and Directrix of a Parabola

✎ Use the information provided to write the vertex form equation of each parabola.

1) $y = x^2 + 8x$

2) $y = x^2 - 6x + 5$

3) $y + 6 = (x + 3)^2$

4) $y = x^2 + 10x + 33$

5) $y = (x + 5)(x + 4)$

6) $\frac{1}{2}(y + 4) = (x - 7)^2$

7) $162 + 731 = -y - 9x^2$

8) $y = x^2 + 16x + 71$

9) Focus: $(-\frac{63}{8}, -7)$, Directrix: $x = -\frac{65}{8}$

10) Focus: $(\frac{107}{12}, -7)$, Directrix: $x = \frac{109}{12}$

11) Opens down or up, and passes through $(-6, -7), (-11, -2),$ and $(-8, 1)$

12) Opens down or up, and passes through $(11, 15), (7, 7),$ and $(4, 22)$

Standard Form of a Circle

✎ **Write the standard form equation of each circle.**

1) $x^2 + y^2 - 8x - 6y + 21 = 0$

2) $y^2 + 2x + x^2 = 24y - 120$

3) $x^2 + y^2 - 2y - 15 = 0$

4) $8x + x^2 - 2y = 64 - y^2$

5) Center: (–5, –6), Radius: 9

6) Center: (–9, –12), Radius: 4

7) Center: (–12, –5), Area: 4π

8) Center: (–11, –14), Area: 16π

9) Center: (–3, 2), Circumference: 2π

10) Center: (15, 14), Circumference: $2\pi\sqrt{15}$

✎ **Identify the center and radius of each. Then sketch the graph.**

11) $(x - 2)^2 + (y + 5)^2 = 10$

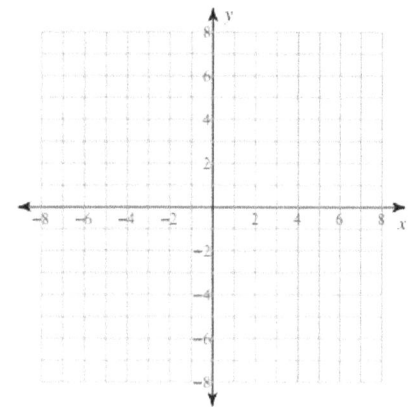

12) $x^2 + (y - 1)^2 = 4$

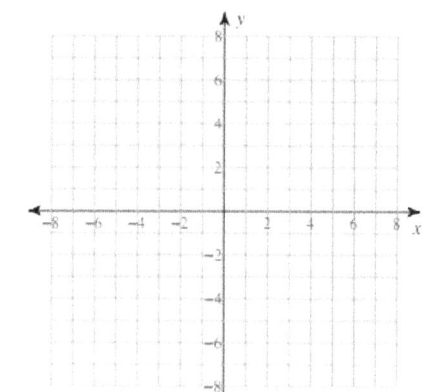

13) $(x - 2)^2 + (y + 6)^2 = 9$

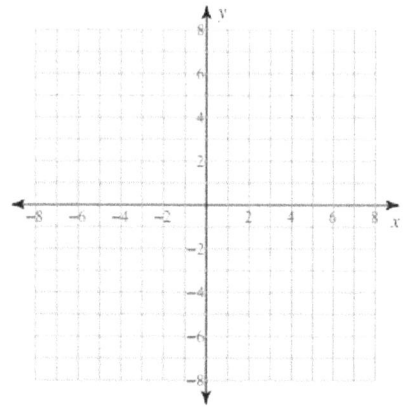

14) $(x + 14)^2 + (y - 5)^2 = 16$

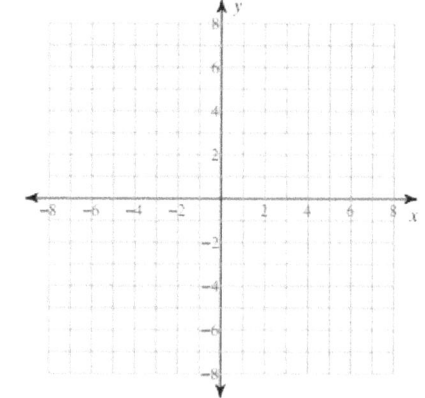

Equation of Each Ellipse

✎ **Use the information provided to write the standard form equation of each ellipse.**

1) Foci: $(2\sqrt{3}, 0), (-2\sqrt{3}, 0)$; Co–vertices: $(0, 2). (0, -2)$

2) Vertices: $(0, 6), (0, -6)$; Co–vertices: $(3, 0). (-3, 0)$

3) Vertices: $(4, 3), (4, -7)$; Co–vertices: $(1, -2). (7, -2)$

4) Foci: $(\sqrt{17}, 0), (-\sqrt{17}, 0)$; Co–vertices: $(9, 0). (-9, 0)$

5) Foci: $(-7, 5 + \sqrt{13}), (-7, 5 - \sqrt{13})$; Co–vertices: $(-1, 5). (-13, 5)$

6) Vertices: $(5, 1), (-1, 1)$; Co–vertices: $(2, 3). (2, -1)$

7) Vertices: $(12, 0), (-12, 0)$; Co–vertices: $(2\sqrt{11}, 0). (-2\sqrt{11}, 0)$

8) Vertices: $(7 + 2\sqrt{35}, -4), (7 - 2\sqrt{35}, -4)$; Co–vertices: $(7, -2). (7, -6)$

9) Center: $(4, 8)$; Vertex: $(4, 8 - \sqrt{170})$; Co–vertex: $(4 - \sqrt{15}, 8)$

10) Center: $(7, -10)$; Vertex: $(-6, -10)$; Co–vertex: $(7, -17)$

✎ **Identify the vertices, co–vertices, foci**

11) $\frac{x^2}{169} + \frac{y^2}{64} = 1$

12) $\frac{x^2}{95} + \frac{y^2}{30} = 1$

13) $\frac{x^2}{36} + \frac{y^2}{16} = 1$

14) $\frac{x^2}{49} + \frac{y^2}{169} = 1$

15) $\frac{(x+5)^2}{81} + \frac{(y-1)^2}{144} = 1$

16) $\frac{(x-3)^2}{49} + \frac{(y-9)^2}{4} = 1$

17) $\frac{x^2}{64} + \frac{(y-8)^2}{9} = 1$

18) $\frac{x^2}{64} + \frac{(y-6)^2}{121} = 1$

Hyperbola in Standard Form

Use the information provided to write the standard form equation of each hyperbola.

1) $-2x^2 + 3y^2 + 4x - 60y + 268 = 0$

2) $-x^2 + y^2 - 18x - 14y - 132 = 0$

3) $-16x^2 + 9y^2 + 32x + 144y - 16 = 0$

4) $9x^2 - 4y^2 - 90x + 32y - 163 = 0$

5) Vertices: (8, 14), (8, –10), Conjugate Axis is 6 units long

6) Vertices: (7, 4), (7, –24), Distance from Center to Focus = $7\sqrt{5}$

7) Vertices: (–5, 22), (–5, –4), Distance from Center to Focus = $\sqrt{218}$

8) Vertices: (0, –1), (–20, –1), Asymptotes: $y = x + 9$, $y = -x - 11$

9) Foci: $(-9, -5 + 9\sqrt{2})$, $(-9, -5 - 9\sqrt{2})$; Conjugate Axis is 18 units long

10) Foci: $(8, -5 + \sqrt{53})$, $(8, -5 - \sqrt{53})$,

    Endpoints of Conjugate Axis: (15, –5), (1, –5)

Identify the vertices, foci, and direction of opening of each.

11) $\dfrac{y^2}{25} - \dfrac{x^2}{16} = 1$

12) $\dfrac{x^2}{121} - \dfrac{y^2}{36} = 1$

13) $\dfrac{x^2}{121} - \dfrac{y^2}{81} = 1$

14) $\dfrac{x^2}{81} - \dfrac{y^2}{4} = 1$

15) $\dfrac{(x+2)^2}{169} - \dfrac{(y+8)^2}{4} = 1$

16) $\dfrac{(y+8)^2}{36} - \dfrac{(y+2)^2}{25} =$

Conic Sections in Standard Form

✍ **Classify each conic section and write its equation in standard form.**

1) $x^2 - 4y^2 + 6x - 8y + 1 = 0$

2) $3x^2 + 3x + y + 79 = 0$

3) $x^2 + y^2 + 4x - 2y - 18 = 0$

4) $-y^2 + x + 8y - 17 = 0$

5) $49x^2 + 9y^2 + 392x + 343 = 0$

6) $-9x^2 + y^2 - 72x - 153 = 0$

7) $-2y^2 + x - 20y - 49 = 0$

8) $-x^2 + 10x + y - 21 = 0$

✍ **Classify each conic section. (Not in Standard Form)**

9) $x^2 + y^2 - 8x + 8y - 4 = 0$

10) $y = 6x^2 - 60x + 149$

11) $x^2 - 4x + 4y^2 - 32y + 32 = 0$

12) $x^2 - 2x - 36y^2 - 360y - 935 = 0$

13) $y = 6x^2 - 60x + 149$

14) $x^2 + y^2 - 8x + 8y - 4 = 0$

15) $x^2 + y^2 + 6x + 10y + 33 = 0$

16) $x^2 - 4x - 36y^2 + 288y - 608 = 0$

17) $9x^2 + 4y^2 + 16y - 128 = 0$

18) $x^2 + 8x - 25y^2 + 50y - 34 = 0$

19) $y = 6x^2 + 60x + 155$

20) $4x^2 + 9y^2 - 54y + 45 = 0$

21) $-9x^2 - 54x + 4y^2 - 40y - 125 = 0$

22) $x^2 - 4x + 4y^2 - 32y + 32 = 0$

Answers of Worksheets – Chapter 12

Equation of a Parabola

1) $x^2 = 8y$
2) $(x - 3)^2 = 8(y - 2)$
3) $(x - 1)^2 = 20(y - 1)$
4) $(x + 1)^2 = 12(y - 2)$
5) $(x - 2)^2 = 8(y - 2)$
6) $x^2 = 8(y - 1)$
7) $(y - 1)^2 = 8(x - 2)$
8) $(y - 1)^2 = 8(x - 2)$
9) $(y - 4)^2 = 16(x + 2)$
10) $(y + 4)^2 = 16x$

Focus, Vertex, and the Directrix of a Parabola

1) $y = (x + 4)^2 - 16$
2) $y = (x - 3)^2 - 4$
3) $y = (x + 3)^2 - 6$
4) $y = (x + 5)^2 + 8$
5) $y = (x + \frac{9}{2})^2 - \frac{1}{4}$
6) $y = 2(x - 7)^2 - 4$
7) $y = -9(x + 9)^2 - 2$
8) $y = (x + 8)^2 + 7$
9) $x = 2(y + 7)^2 - 8$
10) $x = -3(y + 7)^2 + 9$
11) $y = -(x + 9)^2 + 2$
12) $y = (x - 8)^2 + 6$

Standard Form of a Circle

1) $(x - 4)^2 + (y - 3)^2 = 4$
2) $(x + 1)^2 + (y - 12)^2 = 25$
3) $x^2 + (y - 1)^2 = 16$
4) $(x + 4)^2 + (y - 1)^2 = 81$
5) $(x + 5)^2 + (y + 6)^2 = 81$
6) $(x + 9)^2 + (y + 12)^2 = 16$
7) $(x + 12)^2 + (y + 5)^2 = 4$
8) $(x + 11)^2 + (y + 14)^2 = 16$
9) $(x + 3)^2 + (y - 2)^2 = 1$
10) $(x - 15)^2 + (y - 14)^2 = 15$

11) Center: $(2, -5)$, Radius: $\sqrt{10}$

12) Center: $(0, 1)$, Radius: $2\sqrt{26}$

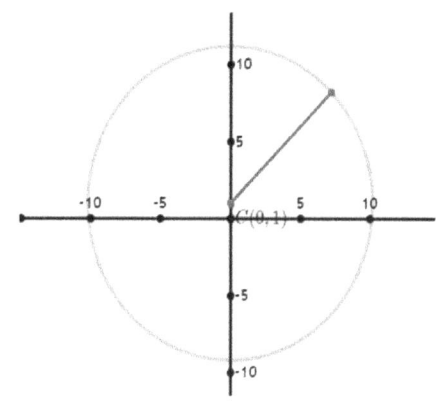

13) Center: (2, −6), Radius: 3

14) Center: (−14, −5), Radius: 4

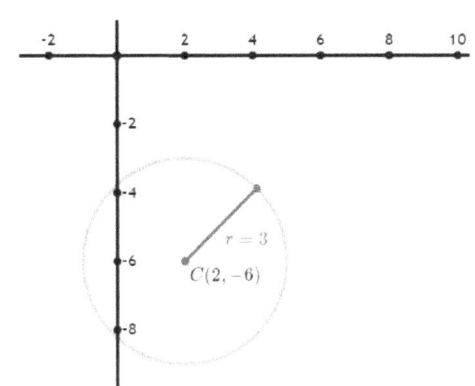
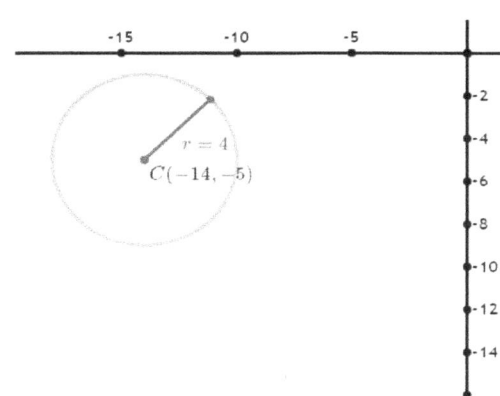

Equation of Each Ellipse

1) $\dfrac{x^2}{16}+\dfrac{y^2}{4}=1$

2) $\dfrac{x^2}{9}+\dfrac{y^2}{36}=1$

3) $\dfrac{(x-4)^2}{9}+\dfrac{(y+2)^2}{25}=1$

4) $\dfrac{x^2}{81}+\dfrac{y^2}{64}=1$

5) $\dfrac{(x+7)^2}{36}+\dfrac{(y-5)^2}{49}=1$

6) $\dfrac{(x-2)^2}{9}+\dfrac{(y-1)^2}{4}=1$

7) $\dfrac{x^2}{144}+\dfrac{y^2}{100}=1$

8) $\dfrac{(x-7)^2}{144}+\dfrac{(y+5)^2}{4}=1$

9) $\dfrac{(x-4)^2}{15}+\dfrac{(y-8)^2}{170}=1$

10) $\dfrac{(x-7)^2}{169}+\dfrac{(y+10)^2}{49}=1$

11) Vertices: (13, 0), (−13, 0); Co–vertices: (0, 8), (0, −8); Foci: ($\sqrt{105}$, 0), (−$\sqrt{105}$, 0)

12) Vertices: ($\sqrt{95}$, 0), (−$\sqrt{95}$, 0); Co–vertices: (0, $\sqrt{30}$), (0, −$\sqrt{30}$); Foci: ($\sqrt{65}$, 0), (−$\sqrt{65}$, 0)

13) Vertices: (6, 0), (−6, 0); Co–vertices: (0, 4), (0, −4); Foci: ($2\sqrt{5}$, 0), (−$2\sqrt{5}$, 0)

14) Vertices: (0, 13), (0, −13); Co–vertices: (7, 0), (−7, 0); Foci: (0, $2\sqrt{30}$), (0, −$2\sqrt{30}$)

15) Vertices: (−5, 13), (−5, −11); Co–vertices: (4, 1), (−14, 1);

   Foci: (−5, $1+3\sqrt{7}$), (−5, $1-3\sqrt{7}$)

16) Vertices: (10, 9), (−4, 9); Co–vertices: (3, 11), (3, 7);

   Foci: ($3+3\sqrt{5}$, 9), ($3-3\sqrt{5}$, 9)

17) Vertices: (8, 8), (−8, 8); Co–vertices: (0, 11), (0, 5);

   Foci: ($\sqrt{55}$, 8), (−$\sqrt{55}$, 8)

18) Vertices: (0, 17), (0, −5); Co–vertices: (8, 6), (−8, 6);

   Foci: (0, $6+\sqrt{57}$), (0, $6-\sqrt{57}$)

CLEP College Algebra Workbook

Hyperbola in Standard Form

1) $\dfrac{(y-10)^2}{10} - \dfrac{(x-1)^2}{15} = 1$

2) $\dfrac{(y-7)^2}{100} - \dfrac{(x+9)^2}{100} = 1$

3) $\dfrac{(y+8)^2}{64} - \dfrac{(x-1)^2}{36} = 1$

4) $\dfrac{(x-5)^2}{36} - \dfrac{(y-4)^2}{81} = 1$

5) $\dfrac{(y-2)^2}{144} - \dfrac{(x-8)^2}{9} = 1$

6) $\dfrac{(y+10)^2}{196} - \dfrac{(x-7)^2}{49} = 1$

7) $\dfrac{(y-9)^2}{196} - \dfrac{(x+5)^2}{49} = 1$

8) $\dfrac{(x+10)^2}{100} - \dfrac{(y+1)^2}{100} = 1$

9) $\dfrac{(y+5)^2}{81} - \dfrac{(x+9)^2}{81} = 1$

10) $\dfrac{(y+5)^2}{4} - \dfrac{(x-8)^2}{49} = 1$

11) Vertices: (0, 5), (0, –5); Foci: (0, $\sqrt{41}$), (0, –$\sqrt{41}$); Opens up/down

12) Vertices: (11, 0), (–11, 0); Foci: ($\sqrt{157}$, 0), (–$\sqrt{157}$, 0); Opens left/right

13) Vertices: (11, 0), (–11, 0); Foci: ($\sqrt{202}$, 0), (–$\sqrt{202}$, 0); Opens left/right

14) Vertices: (9, 0), (–9, 0); Foci: ($\sqrt{85}$, 0), (–$\sqrt{85}$, 0); Opens left/right

15) Vertices: (11, –8), (–15, –8); Foci: (–2 + $\sqrt{173}$, –8), (–2 – $\sqrt{173}$, –8); Opens left/right

16) Vertices: (–2, –2), (–2, –14); Foci: (–2, –8 + $\sqrt{61}$), (–2, –8 – $\sqrt{61}$); Opens up/down

Conic Sections in Standard Form

1) Hyperbola, $\dfrac{(x+3)^2}{4} - (y+1)^2 = 1$

2) Parabola, $y = -3(x+5)^2 - 4$

3) Circle, $(x+2)^2 + (y-1)^2 = 23$

4) Parabola, $x = (y-4)^2 + 1$

5) Ellipse, $\dfrac{(x+4)^2}{9} + \dfrac{y^2}{49} = 1$

6) Hyperbola, $\dfrac{y^2}{9} - (x+4)^2 = 1$

7) Parabola, $x = 2(y+5)^2 - 1$

8) Parabola, $y = (x-5)^2 - 4$

9) Circle

10) Parabola

11) Ellipse

12) Hyperbola

13) Parabola

14) Circle

15) Circle

16) Hyperbola

17) Ellipse

18) Hyperbola

19) Parabola

20) Ellipse

21) Hyperbola

22) Ellipse

Chapter 13:
Statistics and Probability

Topics that you will practice in this chapter:

- ✓ Probability Problems
- ✓ Factorials
- ✓ Combinations and Permutation

Mathematics is no more computation than typing is literature.

— John Allen Paulos

Probability Problems

✎ **Calculate.**

1) A number is chosen at random from 1 to 20. Find the probability of selecting number 8 or smaller numbers. _____

2) Bag A contains 16 red marbles and 6 green marbles. Bag B contains 12 black marbles and 18 orange marbles. What is the probability of selecting a green marble at random from bag A? What is the probability of selecting a black marble at random from Bag B? _____

3) A number is chosen at random from 1 to 25. What is the probability of selecting multiples of 5? _____

4) A card is chosen from a well-shuffled deck of 52 cards. What is the probability that the card will be a queen? _____

5) A number is chosen at random from 1 to 15. What is the probability of selecting a multiple of 4? _____

A spinner, numbered 1–8, is spun once. What is the probability of spinning …?

6) an Odd number? _____   7) a multiple of 2? _____

8) a multiple of 5? _____   9) number 10? _____

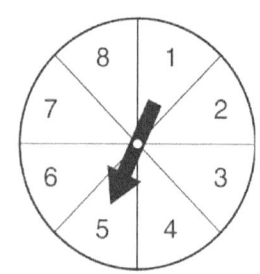

Factorials

✎ **Determine the value for each expression.**

1) $6! + 1! =$

2) $5! + 2! =$

3) $(4!)^2 =$

4) $6! - 3! =$

5) $8! - 4! + 3 =$

6) $3! \times 4 - 12 =$

7) $(3! + 1!)^2 =$

8) $(5! - 4!)^2 =$

9) $(3!\,0!)^2 - 2 =$

10) $\dfrac{8!}{6!} =$

11) $\dfrac{3!}{2!} =$

12) $\dfrac{6!}{5!} =$

13) $\dfrac{21!}{19!} =$

14) $\dfrac{(n-1)!}{(n-3)!} =$

15) $\dfrac{(n+2)!}{(n+1)!} =$

16) $\dfrac{(4+2!)^3}{2!} =$

17) $\dfrac{4n!}{2n!} =$

18) $\dfrac{31!}{29!\,2!} =$

19) $\dfrac{13!}{9!\,3!} =$

20) $\dfrac{6 \times 280!}{3(4 \times 70)!} =$

21) $\dfrac{30!}{31!\,2!} =$

22) $\dfrac{7!\,7!}{8!\,5!} =$

23) $\dfrac{12!\,11!}{9!\,10!} =$

24) $\dfrac{(2 \times 5)!}{1!\,9!} =$

25) $\dfrac{2!(6n-1)!}{(6n)!} =$

26) $\dfrac{n(4n+4)!}{(4n+5)!} =$

27) $\dfrac{(n+1)!(n)}{(n+2)!} =$

Combinations and Permutations

✎ **Calculate the value of each.**

1) 6! = ____

2) 2! × 5! = ____

3) 4! = ____

4) 3! + 5! = ____

5) 7! = ____

6) 9! = ____

7) 3! + 3! = ____

8) 5! − 2! = ____

✎ **Find the answer for each word problems.**

9) Susan is baking cookies. She uses sugar, Vanilla and eggs. How many different orders of ingredients can she try? _____

10) Albert is planning for his vacation. He wants to go to museum, watch a movie, go to the beach, play volleyball and play football. How many ways of ordering are there for him? _____

11) How many 6-digit numbers can be named using the digits 1, 6, 8, 9, and 10 without repetition? _____

12) In how many ways can 4 boys be arranged in a straight line? _____

13) In how many ways can 8 athletes be arranged in a straight line? _____

14) A professor is going to arrange her 5 students in a straight line. In how many ways can she do this? _____

15) How many code symbols can be formed with the letters for the word FRIEND? _____

16) In how many ways a team of 7 basketball players can to choose a captain and co-captain? _____

Answers of Worksheets – Chapter 13

Probability Problems

1) $\frac{2}{5}$
2) $\frac{3}{11}, \frac{2}{5}$
3) $\frac{1}{5}$
4) $\frac{1}{13}$
5) $\frac{1}{5}$
6) $\frac{1}{2}$
7) $\frac{1}{2}$
8) $\frac{1}{8}$
9) 0

Factorials

1) 721
2) 122
3) 576
4) 714
5) 40,299
6) 12
7) 49
8) 9,216
9) 34
10) 56
11) 3
12) 6
13) 420
14) $(n-1)(n-2)$
15) $n+2$
16) 108
17) 2
18) 465
19) 2,860
20) 2
21) $\frac{1}{62}$
22) 5.25
23) 14,520
24) 10
25) $\frac{1}{3n}$
26) $\frac{n}{4n+5}$
27) $\frac{n}{n+2}$

Combinations and Permutations

1) 720
2) 240
3) 24
4) 126
5) 5,040
6) 362,880
7) 12
8) 118
9) 6
10) 120
11) 720
12) 24
13) 40,320
14) 120
15) 720
16) 42

CLEP College Algebra Tests Review

College-Level Examination Program (CLEP) is a series of 33 standardized tests that measures your knowledge of certain subjects. You can earn college credit at thousands of colleges and universities by earning a satisfactory score on a computer based CLEP exam.

The CLEP College Algebra measures your knowledge of math topics generally taught in a one-semester college course in algebra. It contains approximately 60 multiple choice questions to be answered in 90 minutes. Some of these questions are pretest questions that will not be scored. These 60 questions cover: basic algebraic operations; linear and quadratic equations, inequalities, and graphs; algebraic, exponential, and logarithmic functions; and miscellaneous other topics. A scientific calculator is available to students during the entire testing time.

The CLEP College Algebra exam score ranges from 20 to 80 converting to A, B, C, or D based on this score. The letter grade is applied to your college course equivalent.

In this section, there are two complete CLEP College Algebra Tests. Take these tests to see what score you will be able to receive on a real CLEP College Algebra test.

Time to Test

Time to refine your skill with a practice examination

Take practice CLEP College Algebra Tests to simulate the test day experience. After you've finished, score your tests using the answer keys.

Before You Start

- You'll need a pencil, a calculator and a timer to take the test.

- For each question, there are four possible answers. Choose which one is best.

- It's okay to guess. There is no penalty for wrong answers.

- Use the answer sheet provided to record your answers.

- After you've finished the test, review the answer key to see where you went wrong.

Good Luck!

CLEP College Algebra Test Answer Sheets

Remove (or photocopy) these answer sheets and use them to complete the practice tests.

CLEP College Algebra Practice Test

1	A B C D E	21	A B C D E	41	A B C D E
2	A B C D E	22	A B C D E	42	A B C D E
3	A B C D E	23	A B C D E	43	A B C D E
4	A B C D E	24	A B C D E	44	A B C D E
5	A B C D E	25	A B C D E	45	A B C D E
6	A B C D E	26	A B C D E	46	A B C D E
7	A B C D E	27	A B C D E	47	A B C D E
8	A B C D E	28	A B C D E	48	A B C D E
9	A B C D E	29	A B C D E	49	A B C D E
10	A B C D E	30	A B C D E	50	A B C D E
11	A B C D E	31	A B C D E	51	A B C D E
12	A B C D E	32	A B C D E	52	A B C D E
13	A B C D E	33	A B C D E	53	A B C D E
14	A B C D E	34	A B C D E	54	A B C D E
15	A B C D E	35	A B C D E	55	A B C D E
16	A B C D E	36	A B C D E	56	A B C D E
17	A B C D E	37	A B C D E	57	A B C D E
18	A B C D E	38	A B C D E	58	A B C D E
19	A B C D E	39	A B C D E	59	A B C D E
20	A B C D E	40	A B C D E	60	A B C D E

CLEP College Algebra

Practice Test 1

❖ 60 Questions.

❖ Total time for this test: 90 Minutes.

❖ You may use a scientific calculator on this test.

Administered

1) If $f(x) = 3x + 4(x + 1) + 5$ then $f(2x) =$?

   A. $28x + 6$

   B. $14x - 9$

   C. $25x + 4$

   D. $14x + 9$

   E. $14x - 6$

2) If $8x - 8 = 24$, what is the value of $5x - 2$?

   A. 18

   B. 20

   C. 25

   D. 30

   E. 35

3) If $x + y = 0$, $6x - 3y = 27$, which of the following ordered pairs (x, y) satisfies both equations?

   A. $(2, 3)$

   B. $(5, 3)$

   C. $(3, -3)$

   D. $(3, -6)$

   E. $(3, 6)$

4) A line in the xy-plane passes through origin and has a slope of $\frac{1}{4}$. Which of the following points lies on the line?

   A. $(2, 1)$

   B. $(1, 1)$

   C. $(8, 2)$

   D. $(6, 3)$

   E. $(1, 3)$

CLEP College Algebra Workbook

5) If $x \neq -3$ and $x \neq 4$, which of the following is equivalent to $\frac{1}{\frac{1}{x-4}+\frac{1}{x+3}}$?

A. $\frac{(x-4)(x+3)}{(x-4)+(x+3)}$

B. $\frac{(x+3)+(x-4)}{(x+3)(x-4)}$

C. $\frac{(x+3)(x-4)}{(x+3)-(x-4)}$

D. $\frac{(x+3)+(x-4)}{(x+3)-(x-4)}$

E. $\frac{(x+3)(x-4)}{(x+3)(x-4)}$

6) Which of the following is equivalent to $(4n^2 + 3n + 5) - (3n^2 - 6)$?

A. $n + 4n^2$

B. $n^2 - 6$

C. $n^2 + 3n + 11$

D. $n + 10$

E. $n + 11$

7) If $(ax + 2)(bx + 3) = 6x^2 + cx + 6$ for all values of x and $a + b = 5$, what are the two possible values for c?

A. 22, 21

B. 20, 22

C. 13, 24

D. 24, 13

E. 24, 26

$$y < a - x, y > x + b$$

8) In the xy-plane, if $(0, 0)$ is a solution to the system of inequalities above, which of the following relationships between a and b must be true?

A. $a < b$

B. $a > b$

C. $a = b$

D. $a = b + a$

E. $a = 2b$

CLEP College Algebra Workbook

9) Which of the following points lies on the line that goes through the points $(2, 5)$ and $(4, 6)$?

   A. $(6, 9)$

   B. $(9, 6)$

   C. $(6, 8)$

   D. $(6, 7)$

   E. $(6, -7)$

10) Calculate $f(4)$ for the following function f. $f(x) = x^2 - 3x$

    A. 4

    B. 5

    C. 13

    D. 20

    E. 30

11) John buys a pepper plant that is 6 inches tall. With regular watering the plant grows 4 inches a year. Writing John's plant's height as a function of time, what does the y −intercept represent?

    A. The y −intercept represents the rate of grows of the plant which is 6 inches

    B. The y −intercept represents the starting height of 6 inches

    C. The y −intercept represents the rate of growth of plant which is 4 inches per year

    D. There is no y −intercept

12) If $\frac{3}{x} = \frac{9}{x-8}$ what is the value of $\frac{x}{2}$?

    A. 1

    B. 3

    C. -2

    D. 2

    E. 4

13) Which of the following is an equation of a circle in the xy-plane with center $(0, 2)$ and a radius with endpoint $(\frac{4}{3}, 4)$?

A. $(x + 1)^2 + (y - 2)^2 = \frac{52}{9}$

B. $2x^2 + (y + 2)^2 = \frac{52}{9}$

C. $(x - 2)^2 + (y - 2)^2 = \frac{52}{9}$

D. $x^2 + (y - 2)^2 = \frac{52}{9}$

E. $2x^2 + (y - 2)^2 = \frac{52}{9}$

14) What is the equation of the graph?

A. $x^2 + 6x + 5$

B. $x^2 + 2x + 4$

C. $2x^2 - 4x + 4$

D. $2x^2 + 4x + 2$

E. $4x^2 + 4x + 4$

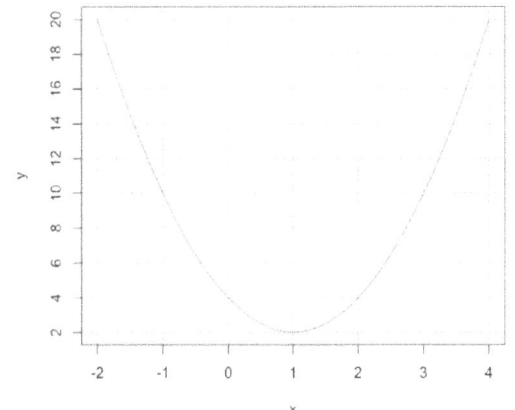

15) What is the solution of the following inequality? $|x - 12| \leq 5$

A. $x \geq 17 \cup x \leq 7$

B. $7 \leq x \leq 17$

C. $x \geq 17$

D. $x \leq 7$

E. Set of real numbers

16) If $4n - 7 \geq 1$, what is the least possible value of $4n + 2$?

A. 3

B. 4

C. 7

D. 9

E. 10

CLEP College Algebra Workbook

17) If $A = \{2, 8, 10, 11\}, B = \{1, 2, 3, 4, 8, 6\}$, and $C = \{8, 7, 9, 10, 12\}$, then which of the following set is $(A \cup B) \cap C$?

   A. $\{1, 2, 3, 4, 8, 6, 11, 10\}$

   B. $\{1, 2, 3, 4, 8, 6, 7, 11, 10, 12\}$

   C. $\{8, 11, 12, 10\}$

   D. $\{8, 10\}$

   E. $\{10\}$

18) The equation $x^2 = 5x - 3$ has how many distinct real solutions?

   A. 0

   B. 1

   C. 2

   D. 3

   E. 4

19) An angle is equal to one fourth of its supplement. What is the measure of that angle?

   A. 20

   B. 36

   C. 45

   D. 60

   E. 90

20) A ladder leans against a wall forming a 60° angle between the ground and the ladder. If the bottom of the ladder is 30 feet away from the wall, how long is the ladder?

   A. 30 feet

   B. 40 feet

   C. 50 feet

   D. 60 feet

   E. 120 feet

CLEP College Algebra Workbook

21) If $x^2 + 4x + r$ factors into $(x + 3)(x + p)$, and r and p are constants, what is the value of r?

   A. 3

   B. 8

   C. 10

   D. 16

   E. 1

22) Simplify.

$$3x^2 + 3y^5 - 2x^2 + z^3 - 2y^2 + 2x^3 - y^5 + 6z^3$$

   A. $x^2 - 2y^2 + y^5 + 7z^3$

   B. $x^2 + 2x^3 - 2y^2 + 2y^5 + 7z^3$

   C. $x^2 + 2x^3 + 3y^5 + 7z^3$

   D. $x^2 + 2x^3 - 2y^2 + 5y^5 + 7z^3$

   E. $x^2 + 2x^3 - 2y^2 + 7z^3$

23) For what value of x is $|x - 4| + 4$ equal to 0?

   A. 1

   B. 2

   C. no value of x

   D. -4

   E. 4

24) Two cars are 240 miles apart. They both drive in a straight line toward each other. If Car A drives at 86 mph and Car B drives at 64 mph, then how many miles apart will they be exactly 40 minutes before they meet?

   A. 20

   B. 45

   C. 60

   D. 80

   E. 100

25) What is the ratio of the minimum value to the maximum value of the following function? $f(x) = -2x + 1 \quad -1 \le x \le 3$

A. $\frac{7}{8}$

B. $-\frac{5}{3}$

C. $-\frac{7}{8}$

D. $\frac{5}{3}$

E. $\frac{6}{7}$

26) If $|a| < 1$, then which of the following is true? $(b > 0)$?

I. $-b < ba < b$

II. $-a < a^2 < a \quad if \ a < 0$

III. $-5 < 2a - 3 < -1$

A. I only

B. II only

C. I and III only

D. III only

E. I, II and III

27) In the triangle below, if the measure of angle A is 35 degrees, then what is the value of y? (figure is NOT drawn to scale)

A. 70

B. 78

C. 86

D. 88

E. 92

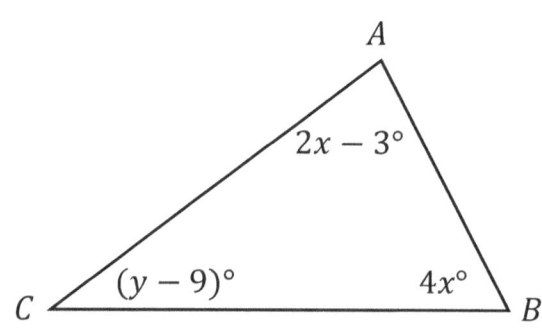

28) 6 years ago, Amy was three times as old as Mike was. If Mike is 12 years old now, how old is Amy?

   A. 4
   B. 8
   C. 12
   D. 14
   E. 24

29) A chemical solution contains 4% alcohol. If there is 32 ml of alcohol, what is the volume of the solution?

   A. 240 ml
   B. 480 ml
   C. 800 ml
   D. 1,200 ml
   E. 2,400 ml

30) If x is a real number, and if $x^3 + 15 = 150$, then x lies between which two consecutive integers?

   A. 1 and 2
   B. 2 and 3
   C. 3 and 4
   D. 4 and 5
   E. 5 and 6

31) If $\frac{4x}{25} = \frac{x-1}{5}$, $x =$?

   A. $\frac{1}{4}$
   B. $\frac{3}{4}$
   C. 3
   D. 5
   E. $\frac{9}{4}$

CLEP College Algebra Workbook

32) If the following equations are true, what is the value of x?

$$a = \sqrt{2}$$

$$4a = \sqrt{4x}$$

A. 2

B. 3

C. 6

D. 8

E. 14

33) If $(x - 6)^3 = 8$ which of the following could be the value of $(x - 6)(x - 5)$?

A. 1

B. 4

C. 6

D. −1

E. −2

34) If $y = nx + 2$, where n is a constant, and when $x = 6$, $y = 20$, what is the value of y when $x = 10$?

A. 10

B. 12

C. 18

D. 32

E. 24

35) Simplify $(-5 + 9i)(2 + 6i)$,

A. $6 - 2i$

B. $60 - 12i$

C. $6 + 2i$

D. $-64 + 12i$

E. $2i$

CLEP College Algebra Workbook

36) Which of the following numbers is NOT a solution of the inequality $2x - 4 \geq 3x - 1$?

   A. -2

   B. -4

   C. -5

   D. -8

   E. -10

37) If $\sqrt{3m - 2} = m$, what is (are) the value(s) of m?

   A. 0

   B. 1

   C. 1, 2

   D. $-1, 2$

   E. $-1, -2$

38) If $\frac{4}{5}y = \frac{4}{3}$, what is the value of y?

   A. $\frac{5}{4}$

   B. $\frac{5}{3}$

   C. $\frac{4}{5}$

   D. $\frac{3}{2}$

   E. $\frac{5}{2}$

39) In 1999, the average worker's income increased $3,000 per year starting from $34,000 annual salary. Which equation represents income greater than average? (I = income, x = number of years after 1999)

   A. $I > 3000x + 34000$

   B. $I > -3000x + 34000$

   C. $I < -3000x + 34000$

   D. $I < 3000x - 34000$

   E. $I < 34,000x + 34000$

40) If the function $g(x)$ has three distinct zeros, which of the following could represent the graph of $g(x)$?

A.
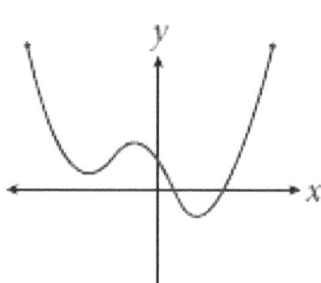

B.
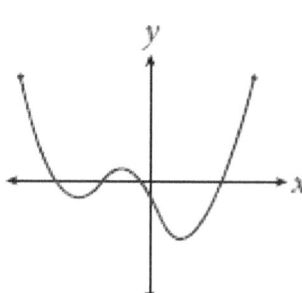

C.
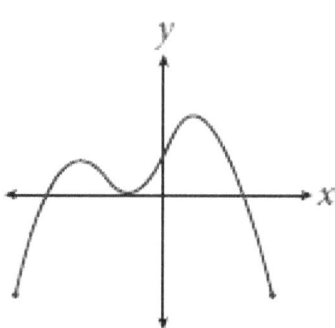

D.
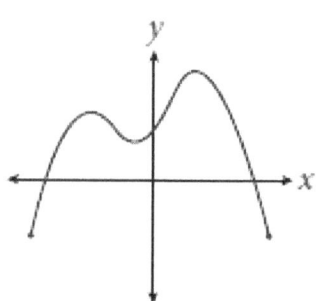

41) If 80% of x equal to 40% of 20, then what is the value of $(x+2)^2$?

A. 12.12

B. 22

C. 23.01

D. 20.25

E. 144

42) In the xy-plane, the point (5,4) and (4,3) are on line A. Which of the following equations of lines is parallel to line A?

A. $y = 2x$

B. $y = 5$

C. $y = \frac{x}{3}$

D. $y = 3x$

E. $y = x$

43) In the following equation when z is divided by 5, what is the effect on x?

$$x = \frac{8y + \frac{r}{r+1}}{\frac{6}{z}}$$

A. x is divided by 2

B. x is divided by 5

C. x does not change

D. x is multiplied by 5

E. x is multiplied by 2

44) A boat sails 40 miles south and then 30 miles east. How far is the boat from its start point?

A. 56 miles

B. 170 miles

C. 80 miles

D. 90 miles

E. 60 miles

45) For what real value of x the equation below is true?

$$x^3 - 5x^2 + 3x - 15 = 0$$

A. 2

B. 4

C. 5

D. 6

E. 8

46) The equation $x^2 = 5x - 6$ has how many distinct real solutions?

A. 0

B. 1

C. 2

D. 3

E. 4

47) If $f(x) = 5^x$ and $g(x) = \log_5 x$, which of the following expressions is equal to

$f(5g(p))$?

A. $5P$

B. 5^p

C. p^5

D. p^9

E. $\frac{p}{5}$

48) The cost of using a car is $0.32 per minutes. Which of the following equations represents the total cost c, in dollars, for h hours of using the car?

A. $c = \frac{60h}{0.32}$

B. $c = \frac{0.32}{60h}$

C. $c = 0.32\,(60h)$

D. $c = 60h + 0.32$

E. $c = 0.32h + 60$

49) Mary's average score after 5 tests is 90. What score on the 6th test would bring Mary's average up to exactly 91?

A. 96

B. 90

C. 98

D. 94

E. 92

50) If $y = 4ab + 3b^3$, what is y when $a = 2$ and $b = 1$?

A. 24

B. 30

C. 36

D. 51

E. 25

51) If $f(x) = 5 + x$ and $g(x) = -x^2 - 2 - 3x$, then find $(g - f)(x)$?

   A. $x^2 - 4x - 7$

   B. $x^2 - 4x + 7$

   C. $-x^2 - 4x + 7$

   D. $-x^2 - 4x - 7$

   E. $-x^2 + 4x - 7$

52) x is $y\%$ of what number?

   A. $\frac{100x}{y}$

   B. $\frac{100y}{x}$

   C. $\frac{x}{100y}$

   D. $\frac{y}{100x}$

   E. $\frac{xy}{100}$

53) In the xy-plane, the line determined by the points $(6, m)$ and $(m, 18)$ passes through the origin. Which of the following could be the value of m?

   A. $\sqrt{6}$

   B. 12

   C. $6\sqrt{3}$

   D. 9

   E. 11

54) A function $g(3) = 5$ and $g(5) = 6$. A function $f(5) = 7$ and $f(6) = 2$. What is the value of $f(g(5))$?

   A. 5

   B. 2

   C. 7

   D. 3

   E. 9

55) Which of the following points lies on the line $2x + 4y = 8$?

   A. (2, 1)   D. (2, 2)

   B. (−1, 3)   E. (2, 8)

   C. (−2, 2)

56) Point A lies on the line with equation $y - 5 = 2(x + 5)$. If the x −coordinate of A is 6, what is the y −coordinate of A?

   A. 14   D. 27

   B. 16   E. 31

   C. 22

$$\frac{c - d}{c} = a$$

57) In the equation above, if c is negative and d is positive, which of the following must be true?

   A. $a < 1$ D. $a < -1$

   B. $a = 0$ E. $a < -2$

   C. $a > 1$

58) If $f(x) = 2x^3 + 4x^2 + 3x$ and $g(x) = -2$, what is the value of $f(g(x))$?

   A. 36   D. 4

   B. 32   E. −1

   C. 24

CLEP College Algebra Workbook

59) If m is a positive integer and $\sqrt{3m+54} = m$, what is the value of m?

   A. 9

   B. 6

   C. 18

   D. 24

   E. 54

60) For what value of x is the function f(x) undefined?

$$f(x) = \frac{1}{(x+5)^2 + 4(x+5) + 4}$$

   A. $x \neq 4$

   B. $x \neq 5$

   C. $x \neq 7$

   D. $x \neq 0$

   E. $x \neq 25$

STOP

This is the End of this Test. You may check your work on this Test if you still have time.

CLEP College Algebra

Practice Test 2

❖ 60 Questions.

❖ Total time for this test: 90 Minutes.

❖ You may use a scientific calculator on this test.

Administered

CLEP College Algebra Workbook

1) In the standard (x, y) coordinate plane, which of the following lines contains the points $(3, -5)$ and $(8, 10)$?

   A. $y = 3x - 10$

   B. $y = \frac{1}{3}x + 13$

   C. $y = -3x + 7$

   D. $y = -\frac{1}{3}x + 10$

   E. $y = 2x - 11$

2) If $f(x) = 2x - 1$ and $g(x) = x^2 - x$, then find $(\frac{f}{g})(x)$.

   A. $\frac{2x-1}{x^2-x}$

   B. $\frac{x-1}{x^2-x}$

   C. $\frac{x-1}{x^2-1}$

   D. $\frac{2x+1}{x^2+x}$

   E. $\frac{x^2-x}{2x-1}$

3) If $\frac{x-3}{4} = N$ and $N = 6$, what is the value of x?

   A. 25

   B. 28

   C. 30

   D. 27

   E. 35

4) Which of the following is equal to $b^{\frac{5}{7}}$?

   A. $\sqrt{b^{\frac{5}{7}}}$

   B. $b^{\frac{5}{7}}$

   C. $\sqrt[7]{b^5}$

   D. $\sqrt[5]{b^5}$

   E. $\sqrt[3]{b^3}$

WWW.MathNotion.Com

CLEP College Algebra Workbook

5) If the interior angles of a quadrilateral are in the ratio 1:2:3:4, what is the measure of the largest angle?

   A. 36°

   B. 72°

   C. 108°

   D. 144°

   E. 180°

6) If the area of a circle is 36 square meters, what is its diameter?

   A. 6π

   B. $6\sqrt{\pi}$

   C. $\frac{6\sqrt{\pi}}{\pi}$

   D. $\frac{6}{\pi}$

   E. $36\pi^2$

7) For $i = \sqrt{-1}$, which of the following is equivalent of $\frac{2+3i}{4-2i}$?

   A. $\frac{3+2i}{5}$

   B. $5+3i$

   C. $\frac{2+16i}{20}$

   D. $\frac{14+16i}{12}$

   E. $\frac{2+16i}{12}$

8) If function is defined as $f(x) = bx^2 + 12$, and b is a constant and $f(2) = 44$. What is the value of $f(1)$?

   A. 20

   B. 35

   C. 60

   D. 65

   E. 75

WWW.MathNotion.Com

9) If $A = \{1, 3, 5, 8, 16, 24\}$ and $B = \{1, 3, 12, 24, 32\}$, how many elements are in $A \cup B$?

   A. 2

   B. 3

   C. 6

   D. 8

   E. 11

10) What is the value of x in the following system of equations?
$$3x + 2y = 3$$
$$6x - 2y = -12$$

   A. -1

   B. 1

   C. -2

   D. 4

   E. 8

11) Calculate $f(3)$ for the function $f(x) = 3x^2 - 6$.

   A. 44

   B. 40

   C. 38

   D. 30

   E. 21

12) In the standard (x, y) coordinate system plane, what is the area of the circle with the following equation? $(x + 2)^2 + (y - 4)^2 = 36$

   A. 8π

   B. 36π

   C. 64π

   D. 64

   E. 128

$$y = x^2 - 5x + 6$$

13) The equation above represents a parabola in the xy-plane. Which of the following equivalent forms of the equation displays the x-intercepts of the parabola as constants or coefficients?

A. $y = x + 3$

B. $y = x(x - 2)$

C. $y = (x + 3)(x + 4)$

D. $y = (x - 3)(x - 4)$

E. $y = (x - 3)(x - 2)$

14) What is the sum of all values of n that satisfies $2n^2 + 12n + 16 = 0$?

A. 6

B. 4

C. −4

D. −6

E. −10

15) Convert 350,000 to scientific notation.

A. 3.50×1000

B. 3.50×10^{-5}

C. 3.5×100

D. 3.5×10^5

E. 3.5×10^4

16) For $i = \sqrt{-1}$, what is the value of $\frac{3+2i}{4+i}$?

A. i

B. $\frac{32i}{5}$

C. $\frac{17-i}{5}$

D. $\frac{14+5i}{17}$

E. $3 + i$

17) The function $g(x)$ is defined by a polynomial. Some values of x and $g(x)$ are shown in the table below. Which of the following must be a factor of $g(x)$?

A. x

B. $x - 1$

C. $x - 3$

D. $x + 1$

E. $x + 3$

x	$g(x)$
0	5
1	4
3	0

18) What is the value of $\frac{9b}{c}$ when $\frac{c}{b} = 3$

A. 8

B. 4

C. 3

D. 1

E. 0

19) Which of the following is equivalent to $\frac{x+(3x)^2+(2x)^3}{x}$?

A. $16x^2 + 27x + 1$

B. $8x^2 + 9x + 1$

C. $16x^2 + 27x$

D. $27x^3 + 16x^2 + 1$

E. $27x^2 + 16x$

20) Which of the following lines is parallel to $6y - 3x = 18$?

A. $y = \frac{1}{2}x + 3$

B. $y = 3x + 5$

C. $y = x - 2$

D. $y = 2x - 1$

E. $y = 4x - 1$

21) If $\frac{a-b}{b} = \frac{9}{10}$, then which of the following must be true?

A. $\frac{a}{b} = \frac{11}{10}$

B. $\frac{a}{b} = \frac{19}{10}$

C. $\frac{a}{b} = \frac{10}{21}$

D. $\frac{a}{b} = \frac{21}{10}$

E. $\frac{a}{b} = \frac{9}{10}$

22) A construction company is building a wall. The company can build 20 cm of the wall per minute. After 30 minutes $\frac{3}{2}$ of the wall is completed. How many meters is the wall?

A. 6

B. 8

C. 4

D. 16

E. 20

23) What is the solution of the following inequality? $|x - 2| \geq 5$

A. $x \geq 7 \cup x \leq -3$

B. $-3 \leq x \leq 7$

C. $x \geq 7$

D. $x \leq -3$

E. Set of real numbers

24) When 5 times the number x is added to 5, the result is 35. What is the result when 3 times x is added to 4?

A. 10

B. 15

C. 22

D. 25

E. 28

25) The average weight of 15 girls in a class is 55 kg and the average weight of 35 boys in the same class is 60 kg. What is the average weight of all the 50 students in that class?

A. 555

B. 58.5

C. 62.68

D. 57.90

E. 58.20

26) If $5h + g = 7h + 6$, what is g in terms of h?

A. $h = 5g - 6$

B. $g = 2h + 6$

C. $h = 4g$

D. $g = h + 1$

E. $g = 5h + 1$

27) If $a - b > 12$ and $a + b < 16$, which of the following pairs could not be the values of a and b?

A. $(9, 0)$

B. $(16, 2)$

C. $(13, 0)$

D. $(14, 1)$

E. $(12, -2)$

28) Simplify.

$$2x^2y^3 + 4x^3y^5 - (5x^2y^3 - 2x^3y^5)$$

A. x^2y^3

B. $6x^2y^3 - 3x^3y^5$

C. $5x^2y^3$

D. $8x^3y^5 - x^2y^3$

E. $3x^5y^8$

29) If $8 + 2x$ is 12 more than 20, what is the value of $6x$?

   A. 40

   B. 55

   C. 62

   D. 72

   E. 88

30) What are the zeroes of the function $f(x) = x^3 + 6x^2 + 8x$?

   A. 0

   B. $-2, -4$

   C. 0, 2, 4

   D. $-1, -4$

   E. $0, -2, -4$

31) If $x + sin^2 a + cos^2 a = 4$, then $x =$?

   A. 2

   B. 5

   C. 3

   D. 6

   E. 4

32) If $\sqrt{7x} = \sqrt{y}$, then $x =$?

   A. $7y$

   B. $\sqrt{\frac{y}{7}}$

   C. $\sqrt{7y}$

   D. y^2

   E. $\frac{y}{7}$

33) If $y = (-4x^3)^2$, which of the following expressions is equal to y?

   A. $-6x^5$

   B. $-6x^6$

   C. $6x^5$

   D. $16x^5$

   E. $16x^6$

CLEP College Algebra Workbook

34) What is the value of the expression $3(x - 2y) + (2 - x)^2$ when $x = 2$ and -1 ?

   A. -3

   B. 15

   C. 12

   D. 30

   E. 25

35) What is the value of x in the following system of equations?

$$6x + 3y = 4$$

$$y = x$$

   A. $x = \frac{4}{9}$

   B. $x = \frac{1}{4}$

   C. $x = \frac{2}{3}$

   D. $x = \frac{4}{3}$

   E. $x = \frac{7}{3}$

36) In a hotel, there are 6 floors and x rooms on each floor. If each room has exactly y chairs, which of the following gives the total number of chairs in the hotel?

   A. $6xy$

   B. $3xy$

   C. $x + y$

   D. $x + 4y$

   E. $2x + 4y$

37) If $\alpha = 2\beta$ and $\beta = 4\gamma$, how many α are equal to 64γ?

   A. 10

   B. 3

   C. 8

   D. 2

   E. 1

$$3x^2 + 5x - 3 \ , \ 2x^2 - 3x + 8$$

38) Which of the following is the sum of the two polynomials shown above?

   A. $6x^2 + 3x + 6$ D. $7x^2 + 4x + 1$

   B. $4x^2 - 7x + 3$ E. $x^2 + 5x + 3$

   C. $5x^2 + 2x + 5$

39) Which of the following is one solution of this equation?

$$x^2 + 4x - 6 = 0$$

   A. $\sqrt{10} - 2$ D. $\sqrt{2} - 1$

   B. $\sqrt{2} + 1$ E. $\sqrt{12}$

   C. $\sqrt{10} + 1$

40) What is the difference in area between an 8 cm by 4 cm rectangle and a circle with diameter of 10 cm? ($\pi = 3$)

   A. 45   D. 3

   B. 43   E. 2

   C. 5

41) Simplify $\frac{3-2i}{-3i}$?

   A. $\frac{2}{3} + i$ D. $\frac{1}{3} + i$

   B. $\frac{2}{3} - i$ E. i

   C. $\frac{1}{3} - i$

x	1	2	3
$g(x)$	-1	-3	-5

42) The table above shows some values of linear function $g(x)$. Which of the following defines $g(x)$?

   A. $g(x) = 2x + 2$

   B. $g(x) = 2x - 2$

   C. $g(x) = -2x + 1$

   D. $g(x) = x + 3$

   E. $g(x) = 2x + 3$

43) Which of the following expressions is equal to $\sqrt{\frac{x^2}{3} + \frac{x^2}{9}}$?

   A. x

   B. $\frac{2x}{3}$

   C. $x\sqrt{x}$

   D. $\frac{x\sqrt{x}}{3}$

   E. $3x$

44) What is the $y-$intercept of the line with the equation $x - 4y = 16$?

   A. 1

   B. -2

   C. 3

   D. -4

   E. 5

45) If $6a - 5 = 13$ what is the value of $5a$?

   A. 10

   B. 15

   C. 40

   D. 35

   E. 20

CLEP College Algebra Workbook

46) If $x \neq 0$ and $x = x^{-4}$, what is the value of x?

   A. -1

   B. 1

   C. 3

   D. 2

   E. 5

47) Which of the following is equal to expression $\frac{6}{x^2} + \frac{5x-3}{x^3}$?

   A. $\frac{6x+1}{x^3}$

   B. $\frac{10x+6}{x^3}$

   C. $\frac{11x-3}{x^3}$

   D. $\frac{13x+2}{x^3}$

   E. $\frac{6x+4}{x^3}$

48) Which of the following is the equation of a quadratic graph with a vertex $(3,-3)$?

   A. $y = 3x^2 - 3$

   B. $y = -3x^2 + 3$

   C. $y = x^2 + 3x - 3$

   D. $y = 4(x-3)^2 - 3$

   E. $y = 4x^2 + 3x - 3$

49) In a coordinate plane, triangle ABC has coordinates: $(-1,6), (-2,5)$, and $(5,8)$. If triangle ABC is reflected over the y-axis, what are the coordinates of the new image?

   A. $(-1,-6), (-2,-5), (-5,-8)$

   B. $(-1,-6), (-2,-5), (5,-8)$

   C. $(1,6), (2,5), (5,8)$

   D. $(-1,6), (-2,5), (5,8)$

   E. $(1,6), (2,5), (-5,8)$

CLEP College Algebra Workbook

50) What is the average of $5x + 1, -2x - 5$ and $9x + 3$?

   A. $2x + 2$

   B. $2x - 2$

   C. $6x + 1$

   D. $3x - \frac{1}{3}$

   E. $x - \frac{1}{3}$

51) If $f(x) = 3x^3 + 3$ and $(x) = \frac{1}{x}$, what is the value of $f(g(x))$?

   A. $\frac{1}{3x^3 + 3}$

   B. $\frac{3}{x^3}$

   C. $\frac{1}{3x}$

   D. $\frac{1}{3x + 3}$

   E. $\frac{3}{x^3} + 3$

52) A plant grows at a linear rate. After five weeks, the plant is 45 cm tall. Which of the following functions represents the relationship between the height (y) of the plant and number of weeks of growth (x)?

   A. $y(x) = 45x + 9$

   B. $y(x) = 9x + 45$

   C. $y(x) = 45x$

   D. $y(x) = 9x$

   E. $y(x) = 4x$

53) A cruise line ship left Port A and traveled 30 miles due west and then 40 miles due north. At this point, what is the shortest distance from the cruise to port A?

   A. 10 miles

   B. 40 miles

   C. 30 miles

   D. 50 miles

   E. 120 miles

WWW.MathNotion.Com

CLEP College Algebra Workbook

54) The length of a rectangle is 4 meters greater than 5 times its width. The perimeter of the rectangle is 32 meters. What is the area of the rectangle?

   A. 16 m²

   B. 24 m²

   C. 64 m²

   D. 28 m²

   E. 90 m²

55) Tickets to a movie cost $12.50 for adults and $7.50 for students. A group of 12 friends purchased tickets for $125. How many student tickets did they buy?

   A. 3

   B. 5

   C. 7

   D. 8

   E. 9

56) If the ratio of $5a$ to $3b$ is $\frac{1}{15}$, what is the ratio of a to b?

   A. 20

   B. 25

   C. $\frac{1}{25}$

   D. $\frac{1}{20}$

   E. $\frac{1}{10}$

57) If $x = 8$, what is the value of y in the following equation?

$$2y = \frac{3x^2}{4} + 2$$

   A. 25

   B. 40

   C. 55

   D. 100

   E. 120

58) What is the value of x in the following equation? $7^x = 2401$

   A. 4

   B. 3

   C. 8

   D. 2

   E. 7

59) If $f(x) = \frac{6x-3}{5}$ and $f^{-1}(x)$, is the inverse of $f(x)$, what is the value of $f^{-1}(3)$?

   A. 5

   B. $\frac{1}{3}$

   C. $\frac{1}{5}$

   D. 3

   E. 15

60) Sara orders a box of pen for $4 per box. A tax of 8.5% is added to the cost of the pens before a flat shipping fee of $8 closest out the transaction. Which of the following represents total cost of p boxes of pens in dollars?

   A. $8p + 4$

   B. $1.085(4p) + 8$

   C. $1.085(8p) + 4$

   D. $4p + 8$

   E. $p + 8$

STOP

This is the End of this Test. You may check your work on this Test if you still have time.

Answers and Explanations

CLEP College Algebra Practice Tests

Answer Key

✳ Now, it's time to review your results to see where you went wrong and what areas you need to improve!

CLEP College Algebra Practice Tests

Practice Test 1						Practice Test 2					
1	D	21	A	41	E	1	A	21	B	41	A
2	A	22	B	42	E	2	A	22	C	42	C
3	C	23	C	43	B	3	D	23	A	43	B
4	A	24	E	44	B	4	C	24	C	44	D
5	C	25	B	45	C	5	D	25	B	45	B
6	C	26	C	46	C	6	C	26	B	46	B
7	C	27	B	47	C	7	C	27	B	47	C
8	B	28	E	48	C	8	A	28	D	48	D
9	D	29	C	49	A	9	D	29	D	49	E
10	A	30	E	50	E	10	C	30	E	50	D
11	B	31	D	51	D	11	E	31	C	51	E
12	C	32	D	52	A	12	B	32	E	52	D
13	D	33	B	53	C	13	E	33	E	53	D
14	C	34	D	54	B	14	D	34	C	54	D
15	B	35	D	55	B	15	D	35	A	55	B
16	E	36	A	56	D	16	D	36	A	56	C
17	D	37	C	57	C	17	C	37	C	57	A
18	C	38	B	58	E	18	C	38	C	58	A
19	B	39	A	59	A	19	B	39	A	59	D
20	D	40	C	60	C	20	A	40	B	60	B

Practice Test 1

Answers and Explanations

1) Answer: D.

If $f(x) = 3x + 4(x + 1) + 5$, then find $f(2x)$ by substituting $2x$ for every x in the function. This gives: $f(2x) = 3(2x) + 4(2x + 1) + 5$,

It simplifies to: $f(2x) = 3(2x) + 4(2x + 1) + 5 = 6x + 8x + 4 + 5 = 14x + 9$

2) Answer: A.

Add 6 both sides of the equation $6x - 6 = 18$ gives $6x = 18 + 6 = 24$.

Dividing each side of the equation $6x = 24$ by 6 gives $x = 4$. Substituting 4 for x in the expression $5x - 2$ gives $5(4) - 2 = 18$.

3) Answer: C.

Method 1: Plugin the values of x and y provided in the options into both equations.

A. $(4, 2)$ $x + y = 0 \to 4 + 2 \neq 0$

B. $(5, 3)$ $x + y = 0 \to 5 + 3 \neq 0$

C. $(3, -3)$ $x + y = 0 \to 3 + (-3) = 0$

D. $(2, -6)$ $x + y = 0 \to 2 + (-6) \neq 0$

E. $(2, 6)$

Only option C is correct.

Method 2: Multiplying each side of $x + y = 0$ by 2 gives $3x + 3y = 0$. Then, adding the corresponding side of $3x + 3y = 0$ and $6x - 3y = 27$ gives $9x = 27$. Dividing each side of $9x = 27$ by 6 gives $x = 3$. Finally, substituting 4 for x in $x + y = 0$, or $y = -3$. Therefore, the solution to the given system of equations is $(3, -3)$.

4) Answer: C.

First, find the equation of the line. All lines through the origin are of the form $y = mx$, so the equation is $y = \frac{1}{4}x$. Of the given choices, only choice C (8,2), satisfies this equation: $y = \frac{1}{4}x \to 2 = \frac{1}{4}(8) = 2$

5) **Answer: A.**

To rewrite $\frac{1}{\frac{1}{x-4}+\frac{1}{x+3}}$, first simplify $\frac{1}{x-4}+\frac{1}{x+3}$.

$$\frac{1}{x-4}+\frac{1}{x+3}=\frac{1(x+3)}{(x-4)(x+3)}+\frac{1(x-4)}{(x+3)(x-4)}=\frac{(x+3)+(x-4)}{(x+3)(x-4)}$$

$$\frac{1}{\frac{1}{x-4}+\frac{1}{x+3}}=\frac{1}{\frac{(x+3)+(x-4)}{(x+3)(x-4)}}=\frac{(x-4)(x+3)}{(x-4)+(x+3)}$$

(Remember, $\frac{1}{\frac{1}{x}}=x$). This result is equivalent to the expression in choice A.

6) **Answer: C.**

$(4n^2 + 3n + 5) - (3n^2 - 6)$; Add like terms together: $4n^2 - 3n^2 = n^2$

$2n$ doesn't have like terms. $5 - (-6) = 11$

Combine these terms into one expression to find the answer: $n^2 + 3n + 11$

7) **Answer: C.**

You can find the possible values of a and b in $(ax + 2)(bx + 3)$ by using the given equation $a + b = 5$ and finding another equation that relates the variables a and b. Since $(ax + 2)(bx + 3) = 6x^2 + cx + 6$, expand the left side of the equation to obtain

$abx^2 + 2bx + 3ax + 6 = 6x^2 + cx + 6$

Since ab is the coefficient of x^2 on the left side of the equation and 10 is the coefficient of x^2 on the right side of the equation, it must be true that $ab = 6$

The coefficient of x on the left side is $2b + 3a$ and the coefficient of x in the right side is c. Then: $2b + 3a = c$, $a + b = 5$, then: $a = 5 - b$

Now, plug in the value of a in the equation $ab = 6$. Then:

$ab = 6 \rightarrow (5-b)b = 6 \rightarrow 5b - b^2 = 6$;

Add $-5b + b^2$ both sides. Then: $b^2 - 5b + 6 = 0$

Solve for b using the factoring method. $b^2 - 5b + 6 = 0 \rightarrow (b-3)(b-2) = 0$

Thus, either $b = 2$ and $a = 3$, or $b = 3$ and $a = 2$. If $b = 2$ and $a = 3$, then

$$2b + 3a = c \rightarrow 2(2) + 3(3) = c \rightarrow c = 13$$

If $b = 3$ and $a = 2$, then, $2b + 3a = c \rightarrow 2(3) + 3(2) = c \rightarrow c = 24$

Therefore, the two possible values for c are 13 and 24.

8) Answer: B.

Since $(0, 0)$ is a solution to the system of inequalities, substituting 0 for x and 0 for y in the given system must result in two true inequalities. After this substitution, $y < a - x$ becomes $0 < a$, and $y > x + b$ becomes $0 > b$. Hence, a is positive and b is negative.

Therefore, $a > b$.

9) Answer: D.

First find the slope of the line using the slope formula. $m = \frac{y_2 - y_1}{x_2 - x_1}$

Substituting in the known information. $(x_1, y_1) = (2, 5), \ (x_2, y_2) = (4, 6)$

$m = \frac{6-5}{4-2} = \frac{1}{2}$. Now the slope to find the equation of the line passing through these points.

$y = mx + b$ Choose one of the points and plug in the values of x and y in the equation to solve for b. Let's choose point $(4, 6)$. Then:

$$y = mx + b \to 5 = \frac{1}{2}(4) + b \to 6 = 2 + b \to b = 6 - 2 = 4$$

The equation of the line is: $y = \frac{1}{2}x + 4$

Now, plug in the points provided in the choices into the equation of the line.

A. $(6, 9)$ $\quad y = \frac{1}{2}x + 4 \to 9 = \frac{1}{2}(6) + 4 \to 9 = 7$ This is NOT true.

B. $(9, 6)$ $\quad y = \frac{1}{2}x + 4 \to 6 = \frac{1}{2}(9) + 4 \to 6 = 8.5$ This is NOT true.

C. $(6, 8)$ $\quad y = \frac{1}{2}x + 4 \to 8 = \frac{1}{2}(6) + 4 \to 8 = 7$ This is NOT true.

D. $(6, 7)$ $\quad y = \frac{1}{2}x + 4 \to 7 = \frac{1}{2}(6) + 4 \to 7 = 7$ This is true!

E. $(6, -7)$ $\quad y = \frac{1}{2}x + 4 \to -7 = \frac{1}{2}(6) + 4 \to -7 = 6$ This is NOT true.

Therefore, the only point from the choices that lies on the line is $(6, 7)$.

10) Answer: A.

The input value is 4. Then: $x = 4$

$f(x) = x^2 - 3x \to f(4) = 4^2 - 3(4) = 16 - 12 = 4$

CLEP College Algebra Workbook

11) Answer: B.

To solve this problem, first recall the equation of a line: $y = mx + b$

Where $m = slope \quad y = y - intercept$

Remember that slope is the rate of change that occurs in a function and that the $y-$intercept is the y value corresponding to $x = 0$.

Since the height of John's plant is 6 inches tall when he gets it. Time (or x) is zero. The plant grows 4 inches per year. Therefore, the rate of change of the plant's height is 4. The $y-$intercept represents the starting height of the plant which is 6 inches.

12) Answer: C.

Multiplying each side of $\frac{3}{x} = \frac{9}{x-8}$ by $x(x - 8)$ gives $3(x - 8) = 9(x)$, distributing the 4 over the values within the parentheses yields $x - 8 = 3x$ or $x = -2$.

Therefore, the value of $\frac{x}{2} = \frac{-4}{2} = -2$.

13) Answer: D.

The equation of a circle can be written as $(x - h)^2 + (y - k)^2 = r^2$ where (h, k) are the coordinates of the center of the circle and r is the radius of the circle. Since the coordinates of the center of the circle are (0, 2), the equation is $x^2 + (y - 2)^2 = r^2$, where r is the radius. The radius of the circle is the distance from the center (0, 2), to the given endpoint of a radius, $\left(\frac{4}{3}, 4\right)$. By the distance formula,

$r^2 = \left(\frac{4}{3} - 0\right)^2 + (4 - 2)^2 = \frac{52}{9}$

Therefore, an equation of the given circle is $x^2 + (y - 2)^2 = \frac{52}{9}$

14) Answer: C.

In order to figure out what the equation of the graph is, fist find the vertex. From the graph we can determine that the vertex is at (1,2). We can use vertex form to solve for the equation of this graph. Recall vertex form, $y = a(x - h)^2 + k$, where h is the x coordinate of the vertex, and k is the y coordinate of the vertex. Plugging in our values, you get $y = a(x - 1)^2 + 2$

To solve for a, we need to pick a point on the graph and plug it into the equation.

Let's pick $(-1, 10)$. $10 = a(-1-1)^2 + 2 \rightarrow 10 = a(-2)^2 + 2 \rightarrow 10 = 4a + 2$

$8 = 4a \rightarrow a = 2$ Now the equation is: $y = 2(x-1)^2 + 2$

Let's expand this, $y = 2(x^2 - 2x + 1) + 2 \rightarrow y = 2x^2 - 4x + 2 + 2 \rightarrow y = 2x^2 - 4x + 4$

The equation in Choice C is the same.

15) Answer: B.

$|x - 12| \leq 5 \rightarrow -5 \leq x - 12 \leq 5 \rightarrow -5 + 12 \leq x - 12 + 12 \leq 5 + 12$

$\rightarrow 7 \leq x \leq 17$

16) Answer: E.

Adding 6 to each side of the inequality $4n - 7 \geq 1$ yields the inequality $4n + 2 \geq 10$. Therefore, the least possible value of $4n + 2$ is 10.

17) Answer: D.

The union of A and B is: $A \cup B = \{1, 2, 3, 4, 8, 6, 11, 10\}$

The intersection of $(A \cup B)$ and C is: $(A \cup B) \cap C = \{8, 10\}$

18) Answer: C.

There can be 0, 1, or 2 solutions to a quadratic equation. In standard form, a quadratic equation is written as: $ax^2 + bx + c = 0$

For the quadratic equation, the expression $b^2 - 4ac$ is called discriminant. If discriminant is positive, there are 2 distinct solutions for the quadratic equation. If discriminant is 0, there is one solution for the quadratic equation and if it is negative the equation does not have any solutions.

To find number of solutions for $x^2 = 5x - 6$ first, rewrite it as $x^2 - 5x + 6 = 0$.

Find the value of the discriminant. $b^2 - 4ac = (-5)^2 - 4(1)(6) = 25 - 24 = 1$

Since the discriminant is positive, the quadratic equation has two distinct solutions.

19) Answer: B.

The sum of supplement angles is 180. Let x be that angle. Therefore, $x + 4x = 180$

$5x = 180$, divide both sides by 5: $x = 36$

20) Answer: D.

The relationship among all sides of special right triangle
$30° − 60° − 90°$ is provided in this triangle:

In this triangle, the opposite side of $30°$ angle is half of the hypotenuse.

Draw the shape of this question:

The ladder is the hypotenuse. Therefore, the ladder is 60 ft.

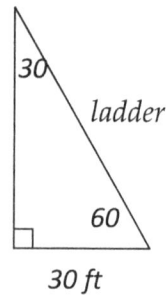

21) Answer: A.

$$x^2 + 4x + r = (x + 3)(x + p) = x^2 + (3 + p)x + 3p$$

On the left side of the equation the coefficient of x is 6 and on the right side of the equation the coefficient of x is $3 + p$.

Thus $3 + p = 4 \rightarrow p = 1$ and $r = 3p = 3(1) = 3$

22) Answer: B.

$$3x^2 + 3y^5 − 2x^2 + z^3 − 2y^2 + 2x^3 − y^5 + 6z^3$$
$$= 3x^2 − 2x^2 + 2x^3 − 2y^2 + 3y^5 − y^5 + z^3 + 6z^3$$
$$= x^2 + 2x^3 − 2y^2 + 2y^5 + 7z^3$$

23) Answer: C.

If the value of $|x − 4| + 4$ is equal to 0, then $|x − 4| + 4 = 0$. Subtracting 4 from both sides of this equation gives $|x − 4| = −4$. The expression $|x − 4|$ on the left side of the equation is the absolute value of $x − 4$, and the absolute value can never be a negative number.

Thus $|x − 4| = −4$ has no solution. Therefore, there are no values for x for which the value of $|x − 4| + 4$ is equal to 0.

24) Answer: E.

The speed of car A is 86 mph and the speed of car B is 64 mph. When both cars drive in a straight line toward each other, the distance between the cars decreases at the rate of 120 miles per hour: 86 + 64 = 150

40 minutes is two third of an hour. Therefore, they will be 100 miles apart 40 minutes before they meet. $\frac{2}{3} \times 150 = 100$

25) Answer: B.

Since $f(x)$ is linear function with a negative slop, then when $x = -1, f(x)$ is maximum and when $x = 3, f(x)$ is minimum. Then the ratio of the minimum value to the maximum value of the function is: $\frac{f(3)}{f(-1)} = \frac{-2(3)+1}{-2(-1)+1} = \frac{-5}{3} = -\frac{5}{3}$

26) Answer: C.

I. $|a| < 1 \to -1 < a < 1$

Multiply all sides by b. Since, $b > 0 \to -b < ba < b$ (it is true!)

II. Since, $-1 < a < 1$, and $a < 0 \to -a > a^2 > a$ (plug in $-\frac{1}{2}$, and check!) (It's false)

III. $-1 < a < 1$, multiply all sides by 2, then: $-2 < 2a < 2$

Subtract 3 from all sides. Then:

$-2 - 3 < 2a - 3 < 2 - 3 \to -5 < 2a - 3 < -1$ (It is true!)

27) Answer: B.

In the figure angle A is labeled $(2x - 3)$ and it measures 10. Thus, $2x - 3 = 35$ and $2x = 38$ or $x = 19$.

That means that angle B, which is labeled $(4x)$, must measure $4 \times 19 = 76$.

Since the three angles of a triangle must add up to 180, $35 + 76 + y - 9 = 180$, then:

$$y + 102 = 180 \to y = 180 - 102 = 78$$

28) Answer: E.

6 years ago, Amy was three times as old as Mike. Mike is 12 years now. Therefore, 6 years ago Mike was 6 years.

6 years ago, Amy was: $A = 3 \times 6 = 18$

Now Amy is 24 years old: $18 + 6 = 24$

29) Answer: C.

4% of the volume of the solution is alcohol. Let x be the volume of the solution.

Then: 4% of $x = 32$ ml $\Rightarrow 0.04 \, x = 32 \Rightarrow x = 32 \div 0.04 = 800$

30) Answer: E.

Solve for x. $\quad x^3 + 15 = 150, \quad x^3 = 135$

Let's review the choices.

A. 1 and 2. $1^3 = 1$ and $2^3 = 8$, 112 is not between these two numbers.

B. 2 and 3. $2^3 = 8$ and $3^3 = 27$, 112 is not between these two numbers.

C. 3 and 4. $3^3 = 27$ and $4^3 = 64$, 112 is not between these two numbers.

D. 4 and 5. $4^3 = 64$ and $5^3 = 125$, 112 is between these two numbers.

E. 5 and 6. $5^3 = 125$ and $6^3 = 216$, 112 is not between these two numbers.

31) Answer: D.

Solve for x. $\frac{4x}{25} = \frac{x-1}{5}$. Multiply the second fraction by 3. $\frac{4x}{25} = \frac{5(x-1)}{5 \times 5}$

Tow denominators are equal. Therefore, the numerators must be equal.

$4x = 5x - 5 \qquad 0 = x - 5 \qquad 5 = x$

32) Answer: D.

Given the two equations, substitute the numerical value of a into the second equation to solve for x. $a = \sqrt{2}$, $4a = \sqrt{4x}$

Substituting the numerical value for a into the equation with x is as follows.

$4(\sqrt{2}) = \sqrt{4x}$, From here, distribute the 4. $4\sqrt{2} = \sqrt{4x}$

Now square both side of the equation. $(4\sqrt{2})^2 = (\sqrt{4x})^2$

Remember to square both terms within the parentheses. Also, recall that squaring a square root sign cancels them out. $4^2\sqrt{2}^2 = 4x$, $16(2) = 4x$, $32 = 4x$, $x = 8$

33) Answer: B.

$(x - 6)^3 = 8 \rightarrow x - 6 = 2 \rightarrow x = 8$

$\rightarrow (x - 6)(x - 5) = (8 - 6)(8 - 5) = (2)(2) = 4$

34) Answer: D.

Substituting 6 for x and 20 for y in $y = nx + 2$ gives $20 = (n)(6) + 2$,

which gives $n = 3$. Hence, $y = 3x + 2$. Therefore, when $x = 10$, the value of y is

$$y = (3)(10) + 2 = 32.$$

35) Answer: D.

We know that: $i = \sqrt{-1} \Rightarrow i^2 = -1$

$(-5 + 9i)(2 + 6i) = -10 - 30i + 18i + 54i^2 = -10 + 12i - 54 = -64 + 12i$

CLEP College Algebra Workbook

36) Answer: A.

Subtracting $2x$ and adding 1 to both sides of $2x - 4 \geq 3x - 1$ gives $-3 \geq x$.

Therefore, x is a solution to $2x - 4 \geq 3x - 1$ if and only if x is less than or equal to -3 and x is NOT a solution to $2x - 4 \geq 3x - 1$ if and only if x is greater than -3.

Of the choices given, only -2 is greater than -3 and, therefore, cannot be a value of x.

37) Answer: C.

First square both sides of the equation to get $3m - 2 = m^2$

Subtracting both sides by $3m - 2$ gives us the equation $m^2 - 3m + 2 = 0$

Here you can solve the quadratic equation by factoring to get $(m - 1)(m - 2) = 0$

For the expression $(m - 1)(m - 2)$ to equal zero, $m = 1$ or $m = 2$

38) Answer: B.

To solve the equation for y, multiply both sides of the equation by the reciprocal of $\frac{4}{5}$, which is $\frac{5}{4}$, this gives $\left(\frac{5}{4}\right) \times \frac{4}{5}y = \frac{4}{3} \times \left(\frac{5}{4}\right)$, which simplifies to $y = \frac{20}{12} = \frac{5}{3}$.

39) Answer: A.

Let x be the number of years. Therefore, $3,000 per year equals $3000x$.

starting from $34,000 annual salary means you should add that amount to $3000x$.

Income more than that is: $I > 3000x + 34000$

40) Answer: C.

A zero of a function corresponds to an x-intercept of the graph of the function in the xy-plane. Therefore, the graph of the function $g(x)$, which has three distinct zeros, must have three x-intercepts. Only the graph in choice C has three x-intercepts.

41) Answer: E.

$0.8x = (0.4) \times 20 \rightarrow x = 10 \rightarrow (x + 2)^2 = (12)^2 = 144$

42) Answer: E.

The slop of line A is: $m = \frac{y_2 - y_1}{x_2 - x_1} = \frac{4 - 3}{5 - 4} = 1$

Parallel lines have the same slope and only choice E ($y = x$) has slope of 1.

43) Answer: B.

Replace z by $z/5$ and simplify.

$$x_1 = \frac{8y + \frac{r}{r+1}}{\frac{6}{\frac{z}{5}}} = \frac{8y + \frac{r}{r+1}}{\frac{5 \times 6}{z}} = \frac{8y + \frac{r}{r+1}}{5 \times \frac{6}{z}} = \frac{1}{5} \times \frac{8y + \frac{r}{r+1}}{\frac{6}{z}} = \frac{x}{5}$$

When z is divided by 5, x is also divided by 5.

44) Answer: B.

Use the information provided in the question to draw the shape.

Use Pythagorean Theorem: $a^2 + b^2 = c^2$

$150^2 + 80^2 = c^2 \Rightarrow 22500 + 6400 = c^2$

$\Rightarrow 28900 = c^2 \Rightarrow c = 170$

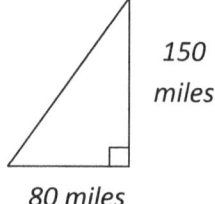

45) Answer: C.

The four-term polynomial expression can be factored completely, by grouping, as follows: $(x^3 - 5x^2) + (3x - 15) = 0$

$x^2(x - 5) + 3(x - 5) = 0 \quad (x - 5)(x^2 + 3) = 0$

By the zero-product property, set each factor of the polynomial equal to 0 and solve each resulting equation for x. This gives $x = 5$ or $x = \pm i\sqrt{3}$, respectively. Because the equation the question asks for the real value of x that satisfies the equation, the correct answer is 5.

46) Answer: C.

To solve a quadratic equation, put it in the $ax^2 + bx + c = 0$ form, factor the left side, and set each factor equal to 0 separately to get the two solutions. To solve $x^2 = 5x - 6$, first, rewrite it as $x^2 - 5x + 6 = 0$. Then factor the left side: $x^2 - 5x + 6 = 0$, $(x - 3)(x - 2) = 0$

$x = 2$ Or $x = 3$, There are two solutions for the equation.

47) Answer: C.

To solve for $f(5g(P))$, first, find $5g(p)$. $g(x) = \log_5 x \rightarrow g(p) = \log_5 p \rightarrow 3g(p) = 5\log_5 p = \log_5 p^5$

Now, find $f(5g(p))$: $f(x) = 5^x \to f(log_5 p^5) = 5^{log_5 p^5}$

Logarithms and exponentials with the same base cancel each other. This is true because logarithms and exponentials are inverse operations. Then: $f(log_5 p^5) = 5^{log_5 p^5} = p^5$

48) Answer: C.

$0.32 per minute to use car. This per-minute rate can be converted to the hourly rate using the conversion 1 hour = 60 minutes, as shown below.

$$\frac{0.32}{minute} \times \frac{60 \, minutes}{1 \, hours} = \frac{\$(0.32 \times 60)}{hour}$$

Thus, the car costs (0.32×60) per hour.

Therefore, the cost c, in dollars, for h hours of use is $c = (0.32 \times 60)h$,

Which is equivalent to $c = 0.32(60h)$

49) Answer: A.

The best way to deal with changing averages is to use the sum. Use the old average to figure out the total of the first 5 scores: Sum of first 5 scores: $(5)(90) = 450$

Use the new average to figure out the total she needs after the 6th score: Sum of 6 score: $(6)(91) = 546$

To get her sum from 450 to 552, Mary needs to score $5546 - 450 = 96$.

50) Answer: E.

$y = 4ab + 3b^3$. Plug in the values of a and b in the equation: $a = 2$ and $b = 1$

$y = 4\,(2)\,(3) + 3\,(1)^3 = 24 + 3(1) = 24 + 1 = 25$

51) Answer: D.

$(g - f)(x) = g(x) - f(x) = (-x^2 - 2 - 3x) - (5 + x)$

$-x^2 - 2 - 3x - 5 - x = -x^2 - 4x - 7$

52) Answer: A.

Let the number be A. Then: $x = y\% \times A$. Solve for A. $x = \frac{y}{100} \times A$

Multiply both sides by $\frac{100}{y}$: $x \times \frac{100}{y} = \frac{y}{100} \times \frac{100}{y} \times A \to A = \frac{100x}{y}$

53) Answer: C.

The line passes through the origin, $(6, m)$ and $(m, 18)$.

Any two of these points can be used to find the slope of the line. Since the line passes through (0, 0) and (6, m), the slope of the line is equal to $\frac{m-0}{6-0} = \frac{m}{6}$. Similarly, since the line passes through (0, 0) and (m, 18), the slope of the line is equal to $\frac{18-0}{m-0} = \frac{18}{m}$. Since each expression gives the slope of the same line, it must be true that $\frac{m}{6} = \frac{18}{m}$

Using cross multiplication gives

$$\frac{m}{6} = \frac{18}{m} \to m^2 = 108 \to m = \pm\sqrt{108} = \pm\sqrt{36 \times 3} = \pm\sqrt{36} \times \sqrt{3} = \pm 6\sqrt{3}$$

54) Answer: B.

It is given that $g(5) = 6$. Therefore, to find the value of $f(g(5))$, then $f(g(5)) = f(6) = 2$

55) Answer: B.

Plug in each pair of number in the equation:

A. (2, 1):   2 (2) + 4 (1) = 8   Bingo!

B. (–1, 3):   2 (–1) + 4 (3) = 10   Nope!

C. (–2, 2):   2 (–2) + 4 (2) = 4   Nope!

D. (2, 2):   2 (2) + 4 (2) = 12   Nope!

E. (2, 8):   2 (2) + 4 (8) = 36   Nope!

56) Answer: D.

Here we can substitute 6 for x in the equation. Thus, $y - 5 = 2(6 + 5)$, $y - 5 = 22$

Adding 5 to both side of the equation: $y = 22 + 5$, $y = 27$

57) Answer: C.

The equation can be rewritten as

$c - d = ac \to$ (divide both sides by c) $1 - \frac{d}{c} = a$, since $c < 0$ and $d > 0$, the value of $-\frac{d}{c}$ is positive. Therefore, 1 plus a positive number is positive. a must be greater than 1. $a > 1$

58) Answer: E.

g(x)=-1, then f(g(x)) = f(-1) = $2(-1)^3 + 4(-1)^2 + 3(-1) = -2 + 4 - 3 = -1$

59) Answer: A.

Squaring both sides of the equation gives $3m + 54 = m^2$

Subtracting both sides by $m + 54$ gives us the equation $m^2 - 3m - 54 = 0$

Here you can solve the quadratic by factoring to get $(m - 9)(m + 6) = 0$

For the expression $(m - 9)(m + 6)$ to equal zero, $m = 9$ or $m = -6$

Since m is a positive integer, 9 is the answer.

60) Answer: C.

The function $f(x)$ is undefined when the denominator of $\frac{1}{(x+5)^2+4(x+5)+4}$ is equal to zero. The expression $(x + 5)^2 + 4(x + 5) + 4$ is a perfect square.

$(x + 5)^2 + 4(x + 5) + 4 = ((x + 5) + 2)^2$ which can be rewritten as $(x + 7)^2$. The expression $(x + 7)^2$ is equal to zero if and only if $x = -7$. Therefore, the value of x for which $f(x)$ is undefined is -7.

Practice Test 2

Answers and Explanations

1) **Answer: A.**

   The equation of a line is: $y = mx + b$, where m is the slope and b is the y-intercept.

   First find the slope: $m = \frac{y_2 - y_1}{x_2 - x_1} = \frac{10 - (-5)}{8 - 3} = \frac{15}{5} = 3$. Then, we have: $y = 3x + b$

   Choose one point and plug in the values of x and y in the equation to solve for b.

   Let's choose the point $(3, -5)$. $y = 3x + b \to -5 = 3(3) + b \to -5 = 9 + b$

   $\to b = -14$; The equation of the line is: $y = 3x - 14$

2) **Answer: A.**

   $(\frac{f}{g})(x) = \frac{f(x)}{g(x)} = \frac{2x - 1}{x^2 - x}$

3) **Answer: D.**

   Since $N = 6$, substitute 6 for N in the equation $\frac{x-3}{4} = N$, which gives $\frac{x-3}{4} = 6$. Multiplying both sides of $\frac{x-3}{4} = 6$ by 5 gives $x - 3 = 24$ and then adding 3 to both sides of $x - 3 = 24$ then, $x = 27$.

4) **Answer: C.**

   $b^{\frac{m}{n}} = \sqrt[n]{b^m}$ For any positive integers m and n. Thus, $b^{\frac{5}{7}} = \sqrt[7]{b^5}$.

5) **Answer: D.**

   The sum of all angles in a quadrilateral is 360 degrees. Let x be the smallest angle in the quadrilateral. Then the angles are: $x, 2x, 3x, 4x$

   $$x + 2x + 3x + 4x = 360 \to 10x = 360 \to x = 36$$

   The angles in the quadrilateral are: $36°, 72°, 108°,$ and $144°$

6) **Answer: C.**

   Formula for the area of a circle is: $A = \pi r^2$. Using 36 for the area of the circle we have: $36 = \pi r^2$. Let's solve for the radius (r).

   $\frac{36}{\pi} = r^2 \to r = \sqrt{\frac{36}{\pi}} = \frac{6}{\sqrt{\pi}} = \frac{6}{\sqrt{\pi}} \times \frac{\sqrt{\pi}}{\sqrt{\pi}} = \frac{6\sqrt{\pi}}{\pi}$

7) Answer: C.

To rewrite $\frac{2+3i}{4-2i}$ in the standard form $a + bi$, multiply the numerator and denominator of $\frac{2+3i}{4-2i}$ by the conjugate, $4 + 2i$. This gives $\left(\frac{2+3i}{4-2i}\right)\left(\frac{4+2i}{4+2i}\right) = \frac{8+4i+12i+6i^2}{4^2-(2i)^2}$. Since $i^2 = -1$, this last fraction can be rewritten as $\frac{8+4i+12i+6(-1)}{16-4(-1)} = \frac{2+16i}{20}$.

8) Answer: A.

First find the value of b, and then find $f(1)$. Since $f(2) = 44$, substituting 2 for x and 44 for $f(x)$ gives $44 = b(2)^2 + 12 = 4b + 12$. Solving this equation gives $b = 8$. Thus: $f(x) = 8x^2 + 12$, $f(1) = 8(1)^2 + 12 \to f(1) = 8 + 12$, $f(1) = 20$

9) Answer: D.

The union of A and B is: $A \cup B = \{1, 3, 5, 8, 12, 16, 24, 32\}$. There are 8 elements in $A \cup B$.

10) Answer: C.

Solving Systems of Equations by Elimination

Multiply the first equation by (–2), then add it to the second equation.

$\begin{matrix} -2(3x+y=3) \\ 6x-4y=-12 \end{matrix} \Rightarrow \begin{matrix} -6x-2y=-6 \\ 6x-4y=-12 \end{matrix} \Rightarrow -6y = -18 \Rightarrow y = 3$

Plug in the value of y into one of the equations and solve for x.

$3x + 2(3) = 3 \Rightarrow 3x + 6 = 3 \Rightarrow 3x = -3 \Rightarrow x = -1$

11) Answer: E.

Identify the input value. Since the function is in the form $f(x)$ and the question asks to calculate $f(3)$, the input value is four. $f(3) \to x = 3$, Using the function, input the desired x value. Now substitute 3 in for every x in the function.

$f(x) = 3x^2 - 6$, $f(3) = 3(3)^2 - 6$, $f(3) = 27 - 6$, $f(3) = 21$

12) Answer: B.

The equation of a circle in standard form is: $(x - h)^2 + (y - k)^2 = r^2$, where r is the radius of the circle. In this circle the radius is 4.

$r^2 = 36 \to r = 6$. $(x + 2)^2 + (y - 4)^2 = 36$

Area of a circle: $A = \pi r^2 = \pi(6)^2 = 36\pi$

13) Answer: E.

The x-intercepts of the parabola represented by $y = x^2 - 5x + 6$ in the xy-plane are the values of x for which y is equal to 0.

The factored form of the equation, $y = (x - 3)(x - 2)$, shows that y equals 0 if and only if $x = 3$ or $x = 2$. Thus, the factored form $y = (x - 3)(x - 2)$, displays the x-intercepts of the parabola as the constants 3 and 4.

14) Answer: D.

The problem asks for the sum of the roots of the quadratic equation $2n^2 + 12n + 16 = 0$. Dividing each side of the equation by 2 gives $n^2 + 6n + 8 = 0$. If the roots of $n^2 + 6n + 8 = 0$ are n_1 and n_2, then the equation can be factored as $n^2 + 6n + 8 = (n - n_1)(n - n_2) = 0$.

Looking at the coefficient of n on each side of $n^2 + 6n + 8 = (n + 4)(n + 2)$ gives $n = -4$ or $n = -2$, then, $-4 + (-2) = -6$

15) Answer: D.

$350000 = 3.5 \times 10^5$

16) Answer: D.

To perform the division $\frac{3+2i}{4+i}$, multiply the numerator and denominator of $\frac{3+2i}{4+1i}$ by the conjugate of the denominator, $5 - i$. This gives $\frac{(3+2i)(4-i)}{(4+i)(4-i)} = \frac{12-3i+8i-2i^2}{4^2-i^2}$. Since $i^2 = -1$, this can be simplified to $\frac{12-3i+8i+2}{16+1} = \frac{14+5i}{17}$

17) Answer: C.

If $x - a$ is a factor of $g(x)$, then $g(a)$ must equal 0. Based on the table $g(3) = 0$. Therefore, $x - 3$ must be a factor of $g(x)$.

18) Answer: C.

To solve this problem first solve the equation for c. $\frac{c}{b} = 3$. Multiply by b on both sides. Then: $b \times \frac{c}{b} = 3 \times b \rightarrow c = 3b$. Now to calculate $\frac{9b}{c}$, substitute the value for c into the denominator and simplify. $\frac{9b}{c} = \frac{9b}{3b} = \frac{9}{3} = \frac{3}{1} = 3$

www.MathNotion.Com

19) Answer: B.

Simplify the numerator. $\frac{x+(3x)^2+(2x)^3}{x} = \frac{x+3^2x^2+2^3x^3}{x} = \frac{x+9x^2+8x^3}{x}$. Pull an x out of each term in the numerator. $\frac{x(1+9x+8x^2)}{x}$. The x in the numerator and the x in the denominator cancel: $1+9x+8x^2 = 8x^2+9x+1$

20) Answer: A.

First write the equation in slope intercept form. Add $2x$ to both sides to get $6y = 3x + 18$. Now divide both sides by 6 to get $y = \frac{1}{2}x + 3$. The slope of this line is $\frac{1}{2}$, so any line that also has a slope of $\frac{1}{2}$ would be parallel to it. Only choice A has a slope of $\frac{1}{2}$.

21) Answer: B.

The equation $\frac{a-b}{b} = \frac{9}{10}$ can be rewritten as $\frac{a}{b} - \frac{b}{b} = \frac{9}{10}$, from which it follows that $\frac{a}{b} - 1 = \frac{9}{10}$, or $\frac{a}{b} = \frac{9}{10} + 1 = \frac{19}{10}$.

22) Answer: C.

The rate of construction company $= \frac{20\ cm}{1\ min} = 20$ cm/min

Height of the wall after 30 minutes $= \frac{20\ cm}{1\ min} \times 30$ min $= 600$ cm

Let x be the height of wall, then $\frac{3}{2}x = 600$ cm $\to x = \frac{2\times 600}{3} \to x = 400$ cm $= 4\ m$

23) Answer: A.

$x - 2 \geq 5 \to x \geq 5 + 2 \to x \geq 7$ Or $x - 2 \leq -5 \to x \leq -5 + 2 \to x \leq -3$

Then, solution is: $x \geq 7 \cup x \leq -3$

24) Answer: C.

When 5 times the number x is added to 5, the result is $5 + 5x$. Since this result is equal to 35, the equation $5 + 5x = 35$ is true. Subtracting 5 from each side of $5 + 5x = 35$ gives $5x = 30$, and then dividing both sides by 5 gives $x = 6$. Therefore, 3 times x added to 4, or $6 + 3x$, is equal to $4 + 3(6) = 22$.

25) Answer: B.

average $= \frac{\text{sum of terms}}{\text{number of terms}}$. The sum of the weight of all girls is: $15 \times 55 = 825$ kg

CLEP College Algebra Workbook

The sum of the weight of all boys is: $35 \times 60 = 2100$ kg. The sum of the weight of all students is: $825 + 2100 = 2925$ kg. average $= \frac{2925}{50} = 58.5$

26) Answer: B.

Fining g in term of h, simply means "solve the equation for g". To solve for g, isolate it on one side of the equation. Since g is on the left-hand side, just keep it there.

Subtract both sides by $5h$. $5h + g - 5h = 7h + 6 - 5h$

And simplifying makes the equation $g = 2h + 6$, which happens to be the answer.

27) Answer: B.

From the choices provided, plugin the values of a and b into both inequalities and check.

A. $(9, -5) \rightarrow a - b = 9 - (-5) = 14 > 12$ and $a + b = 9 + (-5) = 4 < 16$

B. $(16, 2) \rightarrow a - b = 16 - 2 = 14 > 12$ and $a + b = 16 + 2 = 18 > 16$

C. $(13, 0) \rightarrow a - b = 13 - 0 = 13 > 12$ and $a + b = 13 + 0 = 13 < 16$

D. $(14, 1) \rightarrow a - b = 14 - 1 = 13 > 12$ and $a + b = 14 + 1 = 15 < 16$

E. $(12, -2) \rightarrow a - b = 12 - (-2) = 14 > 12$ and $a + b = 12 + (-2) = 12 < 16$

For choice B, 18 is grader than 16. Therefore, choice B does not provide the correct values of a and b.

28) Answer: D.

$2x^2y^3 + 4x^3y^5 - (5x^2y^3 - 2x^3y^5) = 2x^2y^3 - 5x^2y^3 + 4x^3y^5 + 2x^3y^5$
$= 6x^3y^5 - 3x^2y^3$

29) Answer: D.

The description $8 + 2x$ is 12 more than 20 can be written as the equation $8 + 2x = 12 + 20$, which is equivalent to $8 + 2x = 32$. Subtracting 8 from each side of $8 + 2x = 32$ gives $2x = 24$. Since $6x$ is 3 times $2x$, multiplying both sides of $2x = 24$ by 3 gives $6x = 72$

30) Answer: E.

Frist factor the function: $f(x) = x^3 + 6x^2 + 8x = x(x + 4)(x + 2)$

To find the zeros, f(x) should be zero. $f(x) = x(x + 4)(x + 2) = 0$

WWW.MathNotion.Com

Therefore, the zeros are: x = 0, (x + 2) = 0 ⇒ x = −2; (x + 4) = 0 → x = −4

31) **Answer: C.**

$sin^2 a + cos^2 a = 1$, then: $x + 1 = 4$, $x = 3$

32) **Answer: E.**

Solve for x. $\sqrt{7x} = \sqrt{y}$. Square both sides of the equation: $(\sqrt{7x})^2 = (\sqrt{y})^2 \to 7x = y \to x = \frac{y}{7}$

33) **Answer: E.**

$y = (-4x^3)^2 = (-4)^2(x^3)^2 = 16x^6$

34) **Answer: C.**

Plug in the value of x and y. $x = 2$ and $y = -1$

$3(x - 2y) + (2 - x)^2 = 3(2 - 2(-1)) + (2 - 2)^2 = 3(2 + 2) + (0)^2 = 12$

35) **Answer: A.**

Substituting x for y in first equation. $6x + 3y = 4$, $6x + 3(x) = 4$, $9x = 4$

Divide both side of $9x = 4$ by 3 gives $x = \frac{4}{9}$

36) **Answer: A.**

There are 6 floors, x rooms in each floor, and y chairs per room. If you multiply 6 floors by x, there are $6x$ rooms in the hotel. To get the number of chairs in the hotel, multiply $6x$ by y. $6xy$ is the number of chairs in the hotel.

37) **Answer: C.**

If $\beta = 4\gamma$, then multiplying both sides by 16 gives $16\beta = 64\gamma$.

$\alpha = 2\beta$, thus $\alpha = 8\gamma$. Multiply both sides of the equation by 8 gives $8\alpha = 64\gamma$.

38) **Answer: C.**

The sum of the two polynomials is $(3x^2 + 5x - 3) + (2x^2 - 3x + 8)$

This can be rewritten by combining like terms: $(3x^2 + 5x - 3) + (2x^2 - 3x + 8) = (3x^2 + 2x^2) + (5x - 3x) + (-3 + 8) = 5x^2 + 2x + 5$

39) **Answer: A.**

$x_{1,2} = \frac{-b \pm \sqrt{b^2 - 4ac}}{2a}$ $ax^2 + bx + c = 0$

$x^2 + 4x - 6 = 0 \implies$ then: a = 1, b = 4 and c = – 6

$x = \dfrac{-4 + \sqrt{4^2 - 4.1.-6}}{2.1} = \sqrt{10} - 2 \quad x = \dfrac{-4 - \sqrt{4^2 - 4.1.-6}}{2.1} = -2 - \sqrt{10}$

40) Answer: B.

The area of rectangle is: $8 \times 4 = 32$ cm². The area of circle is: $\pi r^2 = \pi \times (\dfrac{10}{2})^2 = 3 \times 25 = 75$ cm². Difference of areas is: $75 - 32 = 43$

41) Answer: A.

To simplify the fraction, multiply both numerator and denominator by i.

$\dfrac{3-2i}{-3i} \times \dfrac{i}{i} = \dfrac{3i - 2i^2}{-3i^2}. \quad i^2 - 1$, Then: $\dfrac{3i - 2i^2}{-3i^2} = \dfrac{3i - 2(-1)}{-3(-1)} = \dfrac{3i + 2}{3} = \dfrac{5i}{3} + \dfrac{2}{3} = i + \dfrac{2}{3}$

42) Answer: C.

Plugin the values of x in the choices provided. The points are $(1, -1), (2, -3),$ and $(3, -5)$

For $(1, -1)$ check the options provided:

A. $g(x) = 2x + 2 \rightarrow -1 = 2(1) + 2 \rightarrow -1 = 4$ This is NOT true.

B. $g(x) = 2x - 2 \rightarrow -1 = 2(1) - 2 = 0$ This is NOT true.

C. $g(x) = -2x + 1 \rightarrow -1 = 2(-1) + 1 \rightarrow -1 = -1$ This is true.

D. $g(x) = x + 3 \rightarrow -1 = 1 + 3 \rightarrow -1 = 4$ This is NOT true.

E. $g(x) = 2x + 2 \rightarrow -1 = 2(1) + 2 = 4$ This is NOT true.

From the choices provided, only choice C is correct.

43) Answer: B.

Simplify the expression.

$\sqrt{\dfrac{x^2}{3} + \dfrac{x^2}{9}} = \sqrt{\dfrac{3x^2}{9} + \dfrac{x^2}{9}} = \sqrt{\dfrac{4x^2}{9}} = \sqrt{\dfrac{4}{9}x^2} = \sqrt{\dfrac{4}{9}} \times \sqrt{x^2} = \dfrac{2}{3} \times x = \dfrac{2x}{3}$

44) Answer: D.

To find the $y-$intercept of a line from its equation, put the equation in slope-intercept form: $x - 4y = 12, \quad -4y = -x + 16, \quad 4y = x - 16, \quad y = \dfrac{1}{4}x - 4$

The $y-$intercept is what comes after the x. Thus, the $y-$intercept of the line is -4.

45) Answer: B.

Adding both side of $6a - 5 = 13$ by 5 gives $6a = 18$

Divide both side of $6a = 18$ by 6 gives $a = 3$, then $5a = 5(3) = 15$

46) Answer: B.

The easiest way to solve this one is to plug the answers into the equation.

When you do this, you will see the only time $x = x^{-4}$ is when $x = 1$ or $x = 0$.

Only $x = 1$ is provided in the choices.

47) Answer: C.

First find a common denominator for both of the fractions in the expression $\frac{6}{x^2} + \frac{5x-3}{x^3}$.

of x^3, we can combine like terms into a single numerator over the denominator:

$$\frac{6x}{x^3} + \frac{5x-3}{x^3} = \frac{(6x) + (5x-3)}{x^3} = \frac{11x - 3}{x^3}$$

48) Answer: D.

Let's find the vertex of each choice provided:

A. $y = 3x^2 - 3$ The vertex is: $(0, -3)$

B. $y = -3x^2 + 3$ The vertex is: $(0, 3)$

C. $y = x^2 + 3x - 3$ The value of x of the vertex in the equation of a quadratic in standard form is: $x = \frac{-b}{2a} = \frac{-3}{2}$

(The standard equation of a quadratic is: $ax^2 + bx + c = 0$)

The value of x in the vertex is 3 not $\frac{-3}{2}$.

D. $y = 4(x - 3)^2 - 3$ Vertex form of a parabola equation is in form of $y = a(x - h)^2 + k$, where (h, k) is the vertex. Then $h = 3$ and $k = -3$. (This is the answer)

E. $y = 4x^2 + 3x - 3$. $x = \frac{-b}{2a} = \frac{-3}{2 \times 8} = -\frac{3}{16}$. The value of x in the vertex is 3 not $-\frac{3}{16}$.

49) Answer: E.

Since the triangle ABC is reflected over the y-axis, then all values of y's of the points don't change and the sign of all x's change. (remember that when a point is reflected over the y-axis, the value of y does not change and when a point is reflected over the x-axis, the value of x does not change). Therefore: (−1,6) changes to (1, 6). (−2, 5) changes to (2, 5). (5, 8) changes to (−5, 8)

50) Answer: D.

To find the average of three numbers even if they're algebraic expressions, add them up and divide by 3. Thus, the average equals:

$$\frac{(5x+1)+(-5x-5)+(9x+3)}{3} = \frac{9x-1}{3} = 3x - \frac{1}{3}$$

51) Answer: E.

$$f(g(x)) = 3 \times \left(\frac{1}{x}\right)^3 + 3 = \frac{3}{x^3} + 3$$

52) Answer: D.

Rate of change (growth or x) is 8 per week. $45 \div 5 = 9$

Since the plant grows at a linear rate, then the relationship between the height (y) of the plant and number of weeks of growth (x) can be written as: $y(x) = 9x$

53) Answer: D.

Use the information provided in the question to draw the shape.

Use Pythagorean Theorem: $a^2 + b^2 = c^2$

$30^2 + 40^2 = c^2 \Rightarrow 900 + 1600 = c^2$

$\Rightarrow 2500 = c^2 \Rightarrow c = 50$

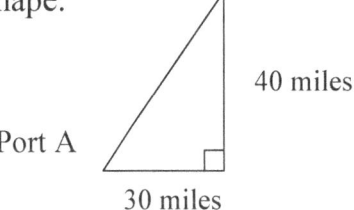

54) Answer: D.

Let L be the length of the rectangular and W be the with of the rectangular. Then, $L = 5W + 4$

The perimeter of the rectangle is 36 meters. Therefore: $2L + 2W = 32 \rightarrow L + W = 16$

Replace the value of L from the first equation into the second equation and solve for W:

$$(5W + 4) + W = 16 \rightarrow 6W + 4 = 16 \rightarrow 6W = 12 \rightarrow W = 2$$

The width of the rectangle is 2 meters and its length is: $L = 5W + 4 = 5(2) + 4 = 14$

The area of the rectangle is: length × width = $2 \times 14 = 28$

CLEP College Algebra Workbook

55) Answer: B.

Let x be the number of adult tickets and y be the number of student tickets. Then:

$$\begin{cases} x + y = 12 \\ 12.50x + 7.50y = 125 \end{cases}$$

Use elimination method to solve this system of equation. Multiply the first equation by -7.5 and add it to the second equation.

$$\begin{cases} -7.5(x + y) = 12 \\ 12.50x + 7.50y = 125 \end{cases} \Rightarrow \begin{cases} -7.5x - 7.5y = -90 \\ 12.50x + 7.50y = 125 \end{cases}$$

$\Rightarrow 5x = 35 \Rightarrow x = 7$

There are 7 adult tickets and 5 student tickets.

56) Answer: C.

Write the ratio of $5a$ to $3b$. $\frac{5a}{3b} = \frac{1}{15}$

Use cross multiplication and then simplify.

$5a \times 15 = 3b \times 1 \to 75a = 3b \to a = \frac{3b}{50} = \frac{b}{25}$

Now, find the ratio of a to b. $\frac{a}{b} = \frac{\frac{b}{25}}{b} \to \frac{b}{25} \div b = \frac{b}{25} \times \frac{1}{b} = \frac{b}{25b} = \frac{1}{25}$

57) Answer: A.

Plug in the value of x in the equation and solve for y.

$2y = \frac{3x^2}{4} + 2 \to 2y = \frac{3(8)^2}{4} + 2 \to 2y = \frac{3(64)}{4} + 2 \to 2y = 48 + 2 = 50$

$2y = 50 \to y = 25$

58) Answer: A.

$2401 = 7^4 \to 7^x = 7^4 \to x = 4$

59) Answer: D.

To solve for the inverse function, first replace $f(x)$ with y. Then, solve the equation for x and after that replace every x with a y and replace every y with an x. Finally, replace y with $f^{-1}(x)$. $f(x) = \frac{6x-3}{5} \Rightarrow y = \frac{6x-3}{5} \Rightarrow 5y = 6x - 3 \Rightarrow 5y + 3 = 6x \Rightarrow$

$\frac{5y+3}{6} = x; f^{-1}(x) = \frac{5x+3}{6} \Rightarrow f^{-1}(3) = \frac{5(3)+3}{6} = \frac{18}{6} = 3$

60) Answer: B.

Since a box of pen costs $3, then $3p$ Represents the cost of p boxes of pen.

Multiplying this number times 1.085 will increase the cost by the 8.5% for tax.

Then add the $8 shipping fee for the total: $1.085(3p) + 8$

"End"

www.ingramcontent.com/pod-product-compliance
Lightning Source LLC
Chambersburg PA
CBHW081108080526
44587CB00021B/3496